TRUTH AND SCIENCE

TRUTH AND SCIENCE

An LDS Perspective

DAVE S. COLLINGRIDGE, PHD

CFI
SPRINGVILLE, UTAH

© 2008 Dave S. Collingridge, PhD

All rights reserved.

No part of this book may be reproduced in any form whatsoever, whether by graphic, visual, electronic, film, microfilm, tape recording, or any other means, without prior written permission of the publisher, except in the case of brief passages embodied in critical reviews and articles.

This is not an official publication of The Church of Jesus Christ of Latter-day Saints. The opinions and views expressed herein belong solely to the author and do not necessarily represent the opinions or views of Cedar Fort, Inc. Permission for the use of sources, graphics, and photos is also solely the responsibility of the author.

ISBN 13: 978-1-59955-082-4

Published by CFI, an imprint of Cedar Fort, Inc., 2373 W. 700 S., Springville, UT, 84663
Distributed by Cedar Fort, Inc., www.cedarfort.com

LIBRARY OF CONGRESS CATALOGING-IN-PUBLICATION DATA

Collingridge, Dave, 1968–
  Truth and science: An LDS Perspective / Dave Collingridge.
      p. cm.
  Includes bibliographical references and index.
  ISBN 978-1-59955-082-4 (alk. paper)
  1. Religion and science. 2. God—Proof. 3. Church of Jesus Christ of
Latter-day Saints—Doctrines. 4. Mormon Church—Doctrines. I. Title. II.
Title: Truth and science.

  BL240.3.C64 2007
  261.5'5—dc22

2007032089

Cover design by Nicole Williams
Cover design © 2008 by Lyle Mortimer
Edited and typeset by Kimiko M. Hammari

Printed in the United States of America

10 9 8 7 6 5 4 3 2 1

Printed on acid-free paper

CONTENTS

Introduction vii

Chapter 1—Spiritual and Secular Truth 1
   "Oh, Say What Is Truth?" 5
   Faith and Reason 8

Chapter 2—Divine Guidance through Ancient Times ... 11
   Sumerians 12
   Egyptians 16
   Mesopotamia 20
   Babylonians 20

Chapter 3—The First Great Illumination 25
   The Golden Age of Greece 25
   Socrates, Plato, Aristotle 28
   Greek Rulers 32

Chapter 4—Early Christianity and the Apostasy 37
   Early Christianity 37
   Apostasy 40
   Dark Ages 45

Chapter 5—The Second Great Illumination 53
   Scientific Revolution and the Restoration 53
   Galileo Galilei 56
   René Descartes 61
   Isaac Newton 64
   Conclusion 68

Chapter 6—Dreams and Inspirations **71**
    The Neuron... 72
    Penicillin 74
    Organic Chemistry... 75
    Malaria 76
    The Periodic Table 78
    Lower Mortality from Disease 80

Chapter 7—Skeptics and Believers **83**
    Theism... 83
    The Enlightenment... 84
    Deism 85
    Agnosticism and Atheism.. 87
    Secularism 91

Chapter 8—Gospel Truth and Science...................... **95**
    The Goal of Science... 97
    Gospel Truth and Assumptions of Naturalism 99
    Conclusion...105

Chapter 9—Bringing Gospel Truth to Science **107**
    Reconciliation 107
    Solving Theological and Philosophical Concerns 109
    Conclusion.. 114

Chapter 10—Gospel Truth and Quantum Mechanics .. **115**
    Quantum Non-Locality 115
    Randomness 117
    Wave-Particle Duality 118

Chapter 11—Coming to the Knowledge of the Truth ... **123**
    "Never Coming to the Knowledge of the Truth"......... 123
    Qualifying for the Enlightening Power of the
        Spirit of the Lord.. 125

Index..**129**

About the Author..**133**

INTRODUCTION

*We should be a people of profound learning pertaining
to the things of the world.*
—Brigham Young

The Light of Christ, Spirit of the Lord, or Light of the Lord, as it is sometimes called, is a divine medium that emanates from the presence of God and fills the immensity of space. It enables God to govern the universe from His celestial home on high and makes possible His omniscience, which is knowing at an instant everything that is happening throughout His creations. The Light of the Lord is also the source of God's omnipresence, making it possible for Him to be everywhere, in and throughout all His creations. It is the source of human conscience as well, the moral compass that teaches us right from wrong and entices us to do good continually.

The Light of Christ is also a source of truth. The Lord uses it to bless mankind with knowledge. In its capacity as a revealer of knowledge, it is often called the Spirit of Truth. The knowledge the Spirit of Truth provides can be divided into two categories, spiritual and secular. These two types of truth differ in terms of the domains in which they were received. Knowledge received in a spiritual or religious domain is often referred to as spiritual, while knowledge discovered outside of religious contexts is often referred to as secular. Of course, at a fundamental level there is no difference between the two; they both contain truth. Their sameness in this regard was affirmed by the Prophet Joseph Smith, who testified that all truth will eventually be circumscribed into one whole.[1] Nevertheless, distinguishing between spiritual and secular truth is helpful when discussing knowledge received in spiritual and secular domains, as is done in this book.

Latter-day Saints are familiar with the notion of the Spirit of the Lord providing spiritual knowledge, most notably in the form of prophetic and personal revelation. However, we are less familiar with the secular knowledge that comes from the Spirit of the Lord. The purpose of this book is to familiarize Latter-day Saints with the secular enlightening power of the Light of Christ. Specifically, this book discusses the ways in which the Lord blesses humanity with secular knowledge through the power of His Spirit. It illustrates that His Spirit is the main source of mankind's scientific knowledge, that it has inspired mankind since the dawn of time, and that without its enlightening power, humanity would flounder in ignorance and darkness.

The Lord has instructed us to seek learning and wisdom by study and faith (D&C 88:118). This instruction refers to spiritual *and* secular knowledge, for He has counseled us to become "acquainted with all good books and with languages, tongues, and people" (D&C 90:15). The prophet Brigham Young similarly urged Latter-day Saints to "read good books, and extract from them wisdom and understanding as much as you possibly can."[2]

Why has the Lord instructed us to seek after secular wisdom and knowledge? Brigham Young provided an answer when he said, "The truth and sound doctrine possessed by the sectarian world, and they have a great deal, all belong to [His] church."[3] He added that Mormonism "embraces all truth, wherever found, in all the works of God and man that are visible or invisible to mortal eye."[4] Thus he encouraged members of the Church "to gather up all the truths in the world pertaining to life and salvation, to the Gospel we preach, to mechanism of every kind, to the sciences, and to philosophy, wherever it may be found in every nation, kindred, tongue, and people and bring it to Zion."[5]

Gathering in valuable knowledge from around the world and bringing it to Zion is not an easy task given that many falsehoods abound. Falsehoods are often hidden under the cloak of scientific fact, theory, and authority. Take evolution, for example. Evolution contains facts in the form of fossil records, biological similarities across species, and micro-evolutionary changes within species. All too often these facts are presented as evidence "proving" macro-evolutionary change from one life form into another. Yet such facts are used to build an inductive generalization, which means that they are used to build the theory of evolution from a common ancestor, not to "prove" the theory. What is lacking is experimental evidence confirming the theory. No one has yet deduced a hypothesis from the theory of macroevolution, tested it in a controlled environment, and confirmed macro-evolutionary change across life forms. Even still, evolutionists continue to assert that their inductive evidence proves that human beings evolved from simple molecules. This is just one instance of falsehoods being perpetuated by well-intentioned scholars who have been duped by "false spirits which have gone forth in the earth, deceiving the world" (D&C 50:2).

Who among us is at risk for falling prey to similar false spirits? Latter-day Saints who are not anchored in gospel teachings may become ensnared by false spirits, lose their faith, and apostatize. Even those with firm testimonies may become ensnared by false spirits and end up mingling scientific falsehoods with gospel teachings. Joseph F. Smith observed that members of the Church "who make but feeble effort, if indeed any at all, to better themselves by reading and study"[6] are most likely to unwittingly combine falsehoods and gospel teachings.

What, then, can we do to avoid becoming ensnared by false spirits? As Joseph F. Smith suggested, we should improve our understanding by studying the relationship between gospel and secular beliefs. Additionally, we should learn about the enlightening power of the Spirit of the Lord and seek to obtain its guidance. Brigham Young taught that our efforts to obtain secular knowledge should be "aided by the Spirit of God,"[7] and added that when we follow the Spirit of the Lord, we not only become more receptive to its guiding influence, but we are also better able to discern between pure knowledge and the false principles of man.[8] By exploring the relationship between secular and gospel beliefs and uncovering the principles upon which the Spirit of the Lord operates, this book will help Latter-day Saints to recognize the enlightening power of the Spirit of the Lord and avoid falsehoods as they fulfill their mandate to bring truth to Zion.

This book also addresses the often intriguing relationship between scientific progress and the gospel. For instance, the reader will discover how the Lord has blessed mankind with a major secular illumination in ancient Greece that laid the groundwork for modern science. During that illumination, the Lord poured out His Spirit upon Greek scholars such as Thales, Socrates, and Aristotle. Their inspired works were instrumental in the eventual overthrow of mythological worship, and prepared the gentiles for the gospel of Jesus Christ which arrived centuries later.

We will also take a look at the connection between the spiritual apostasy and the Dark Ages, and the Restoration of the gospel and scientific progress in the latter days. The rapid scientific progress that we are witnessing today is a direct result of a second illumination that followed the Dark Ages. During this era of illumination, the Lord poured out His Spirit upon scholars such as Nicolaus Copernicus, Isaac Newton, and Galileo Galilei. Their contributions were instrumental in bringing to pass the purposes of God by preparing the inhabitants of the earth for the Restoration of the gospel.

This book also discusses scholarly threats to our basic religious beliefs, refutes atheistic dogma, and explains why, more than ever before, principles of righteousness are needed in secular scholarship. This discussion provides information that will help protect Latter-day Saints from intellectual falsehoods and preserve Christian values in an increasingly secular society.

Additionally, this book contains accounts of marvelous scientific discoveries made possible by the Spirit of the Lord. I have attempted to "make known his wonderful works among the people" (D&C 65:4) by presenting these little-known stories. They demonstrate God's wisdom and love for mankind, and testify that He "doth prosper his people . . . in all manner of precious things of every kind and art; [thus] sparing their lives" (Helaman 12:2).

The Latter-day Saint perspective on secular scholarship contained in this work does not involve revisionism in the sense of distorting historical facts, discoveries, and philosophies. Rather, it is an interpretation of historical facts, discoveries, and philosophies from the perspective of truth found in the restored gospel. Members of The Church of Jesus Christ of Latter-day Saints should find this careful mix of the religious and secular very informative.

Few concepts are entirely original when it comes to scholarship. Most would agree that good ideas are usually built on other good ideas. For this reason I wish to acknowledge my teachers and the numerous authors who have enlightened my understanding. Without their contributions, this work would not have been possible. Finally, I am responsible for the presentation of the ideas contained herein. These ideas are not intended to represent the formal views and doctrines of The Church of Jesus Christ of Latter-day Saints.

Notes

1. See Victor L. Ludlow, *Principles and Practices of the Restored Gospel* (Salt Lake City: Deseret Book Co., 1992), 139.
2. Brigham Young, *Discourses of Brigham Young.* Selected and arranged by John A. Widtsoe (Salt Lake City: Deseret Book Co., 1954), 248.
3. Ibid., 10:251.
4. Ibid., 10:251.
5. Ibid., 248.
6. Joseph F. Smith, *Gospel Doctrine: Selections from the Sermons and Writings of Joseph F. Smith.* Compiled by John A. Widtsoe (Salt Lake City: Deseret Book Co., 1939), 373.
7. Ibid.
8. Ibid., 436.

ONE

SPIRITUAL AND SECULAR TRUTH

I like the truth because it is true, because it is lovely and delightful, because it is so glorious in its nature, and so worthy the admiration, faith and consideration of all intelligent beings in heaven or on the earth.

—Brigham Young

People generally believe that religious and secular scholarship exist in different domains, or that they are concerned with different issues. Galileo recognized this difference when he stated that the purpose of religion is to "teach us how one goes to heaven, not how heaven goes."[1] To a certain extent this is a healthy distinction. History suggests that secular endeavors like science operate best when they are free from theocratic control. Yet, at the same time, Latter-day Saints have a mandate to seek learning by study and prayer. This mandate inevitably leads to an integration of spiritual and secular beliefs in the LDS community.

We also have a mandate to bring truth to the world. The Lord has instructed us to share the doctrines of salvation with others through missionary efforts. But our mandate to spread truth need not stop at the doctrines of salvation; we have much knowledge about the earth, heavens, and humanity that can benefit secular scholarship. We should share this knowledge with the world. Moreover, when secular truth claims arise that are related to gospel teachings, we should lean on gospel teachings as a sort of gold standard by which to evaluate the accuracy of secular claims. Claims that agree with gospel truths may be considered accurate, while those that clearly oppose gospel teachings may be judged false.

There are many ways in which the gospel truths we possess can enrich secular scholarship. Imagine how much more healthy and joyous our society would be if it had heeded the warnings of the Word of Wisdom long ago. We would have avoided much of the heartbreak, sorrow, and ill health typically associated with tobacco use and alcohol consumption. The same can be said for drug abuse and poor eating habits. These vices are taking a toll on our society, yet long ago the Lord warned us about "evils and designs" that would exist in the latter days, and He exhorted us to shun them. But the world did not listen. Only now society is beginning to recognize the wisdom in the Word of Wisdom, after much heartbreak and sorrow. Orson Pratt said that by sharing the truths of the gospel with the outside world, we can put secular endeavors "upon the solid foundations of everlasting truth," and by so doing, help secular knowledge move "higher and still higher, until [it is] crowned with the glory and presence of Him, who is Eternal."[2] This is a lofty goal to which all scholarly minded Latter-day Saints should aspire.

Another way that Latter-day Saints may integrate gospel and secular beliefs is by applying secular truth to our understanding of the gospel. This approach, which Brigham Young called "bringing truth to Zion," must be done in the proper spirit. It must be done with the purpose of enriching lives, strengthening testimonies, and improving our understanding of the gospel. For instance, a forum speaker may mention the complexity and efficiency of the human immune system to illustrate the Lord's wisdom and love for mankind, or talk about the awe-inspiring view of thousands of stars through a telescope to strengthen people's convictions of the Lord's omnipotence.

If bringing truth to Zion is not done in the right spirit, we may unwittingly skew our understanding of the plain and simple truths of the gospel. Latter-day Saints who endeavor to bring secular knowledge to Zion should remember that the guidelines for righteous living and eternal salvation are wholly contained within the gospel; they come from priesthood authority, the Holy Scriptures, and answers to personal prayer, to name a few. Guidelines for righteous living rarely, if ever, come from laboratory experiments, philosophizing, and secular scholarship. Secular learning should never replace our study of the gospel; it should only supplement our understanding and appreciation of gospel principles.

Brigham Young stressed the importance of staying grounded in the gospel while bringing truth to Zion when he taught:

> There are a great many branches of education: some go to college to learn languages, some to study law, some to study physics, and some to study astronomy and various other branches of science. We want every branch of science taught in this place that is taught in the world. But our favorite study is that branch which particularly belongs to the Elders of Israel—namely, theology. Every Elder should become a profound theologian—should understand this branch better than all the world.[3]

The scholarly endeavors of the Prophet Joseph Smith provide a good example of balancing gospel and secular learning. By continually studying the word of God and preparing himself to receive revelation from the Lord, he placed the acquisition of gospel truth above all other types of learning.

Notwithstanding his theological studies, Joseph Smith also actively pursued secular learning through his study of ancient languages such as Greek, Hebrew, and Latin. He studied these languages so that he could better understand the Bible and "read the Scriptures in the language in which they were given."[4] (The oldest biblical manuscripts were written in these three languages.) He wrote, "My soul delights in reading the word of the Lord in the original, and I am determined to pursue the study of the languages, until I shall become master of them, if I am permitted to live long enough. At any rate, so long as I do live, I am determined to make this my object."[5] He hoped that a better understanding of the Bible would enable him to more fully glorify God and hasten the work of the Lord. He said, "O may God give me learning, even language; and endue me with qualifications to magnify His name while I live"[6]

I do not think that he lived long enough to fully grasp Greek, Hebrew, and Latin, yet what he learned certainly helped him glorify the name of the Lord. Have you ever wondered how he formed such eloquent prose given his lack of formal education? Expressions like "I saw a pillar of light exactly over my head, above the brightness of the sun, which descended gradually until it fell upon me," and "I saw two Personages whose brightness and glory defy all description" are not typical of an uneducated nineteenth century farmer. We never grow tired of reading or hearing these words because of their rich descriptive power. They are forever etched in the Latter-day Saint psyche. Without a doubt, the Prophet Joseph Smith's study of languages enabled him to create beautiful expressions that continue to magnify the name of the Lord.

Joseph Smith's emphasis on scholarship had a lasting influence on the Church. He emphasized the importance of education when he established the school of the prophets in Kirkland, a University in Nauvoo, and the Seventies' Hall of Science. His efforts to promote secular learning encouraged many members of the Church to integrate scholarly knowledge and gospel teachings. Latter-day Saint scholars like Orson Pratt, James E. Talmage, John A. Widtsoe, and B. H. Roberts were no doubt influenced by Joseph Smith's interest in religious and secular scholarship.

Joseph Smith's legacy of learning continues today. Daniel H. Ludlow, editor of the *Encyclopedia of Mormonism*, observed that interest in the relationship between science and Mormonism "continues and is presently sustained within the larger Latter-day Saint culture."[7] Indeed, Mormon scholars and everyday members of the Church are continuing Joseph Smith's efforts to enrich our understanding of the gospel by searching for secular truth and bringing it to Zion.

I too have been influenced by the Prophet Joseph Smith's legacy of combining secular and spiritual knowledge. I developed a curiosity in the relationship between the gospel and science at a very young age. While in a ninth grade science class, my teacher announced that scientific discovery began moving at a very fast pace around 1820. Sitting there, I thought, "That's when the First Vision occurred!" At the time I wondered if the similar time frame for the Restoration and rapid scientific progress was a coincidence. I did not think so. I believed then, as I do now, that there is a meaningful relationship between the two. For the next sixteen years I rarely gave this issue serious consideration, but that all changed when I heard a BYU devotional talk by President Hinckley in 1999.

On November 30, 1999, while I was a young graduate student at Brigham Young University in Provo, President Gordon B. Hinckley gave a devotional address titled "Keep the Chain Unbroken." Commenting on the wonderful scientific and technological discoveries of the twentieth century, he declared,

> As we close this great and remarkable century, I stand in awe of the blessings we have. I have now lived through 90 years of this century. When I think of the wonders that have come to pass in my lifetime—more than during the rest of human history together—I stand in reverence and gratitude. . . . It is all so miraculous and wonderful. . . . And with all of this there has been the restoration of the pure gospel of Jesus Christ.[8]

President Hinckley's comments linking the Restoration of the gospel with rapid scientific discovery rekindled my interest in the relationship between the gospel and science. Fortunately, the psychology department's Theory and Philosophy program that I was enrolled in provided ample opportunity for me to explore this relationship. My interests largely centered on the history and philosophy of science. As I studied these topics, my awareness of the gospel's potential to enlighten secular scholarship increased. Whenever I encountered problems regarding truth that confounded secular scholarship, I usually found a satisfactory answer by applying the doctrines of the restored gospel. The principles of the gospel also proved invaluable in understanding events and progress in the history of science.

The chance to converse with others about the relationship between science and religion is one of the strengths of BYU. In harmony with Joseph Smith's legacy of bringing truth to Zion, students are encouraged to integrate religious and secular knowledge. Contrast this with liberal arts universities where discussion about the relationship between religious and scientific knowledge is generally avoided. My experience at two liberal arts schools suggests that these schools embrace, at least implicitly, an attitude that "religious issues are not scientific, so we don't talk about those kinds of things here." This attitude is unfortunate, for as many members of the Church have discovered, the gospel of Jesus Christ has much to offer secular pursuits, especially with regard to understanding truth.

"OH, SAY WHAT IS TRUTH?"

George Q. Cannon stated, "Too great a value cannot be placed upon the possession of the truth. It is indeed beyond estimate."[9] This sentiment is similarly expressed by the words of a familiar hymn which declares that truth is "the brightest prize to which mortals or Gods can aspire." As Latter-day Saints, let us then heed the counsel of that hymn to "Go search in the depths where it glittering lies, or ascend in pursuit to the loftiest skies: 'Tis an aim for the noblest desire."[10] And when we have found honorable truth, let's bring it to Zion, for as Brigham Young declared, the gospel "embraces all truth, wherever found, in all the works of God and man that are visible or invisible to mortal eye."[11]

To fulfill our mandate to find truth and bring it to Zion, we need to have an accurate understanding of the essence of truth. "What is truth?" is a question philosophers have grappled with for millennia. It is a question that Pilate seemed to struggle with during the most important "trial" in the history of the world. Answering this question is not as simple as it might seem, for there are different concepts of truth. Unfortunately, multiple concepts of truth tend to create divisiveness over what is real and indecision about whether we can know anything for certain.

By way of example, let's consider the relative concept of truth, which states that what is true for one person may not be true for another. We see examples of this whenever two individuals perceive the same event in different ways or experience the same thing differently. People may have different perceptions that others cannot deny. What right do we have to say that one perception is right and that another is wrong? Therefore there must be multiple subjective truths, relative to people's perceptions. Such notions are gaining popularity in our society, fueling the moral relativism that seems so prevalent these days. Increasingly, traditional concepts of right versus wrong are being replaced by a mentality of "If it feels good, do it." In addition, moral relativism is leading people to call "evil good, and good evil" in order to excuse their sins.

There are also logical concerns with the relativistic concept of truth. Most notably, when relativism is carried to its logical conclusion, it leads to two self-refuting doctrines known as nihilism and skepticism. Nihilism is the belief that because people have their own perceptions of what is real, there are no absolute truths, so "anything goes" as far as truth claims are concerned. The problem with this doctrine is that if absolute truth does not exist, as nihilism supposes, then the absolute certainty of nihilism cannot exist. Skepticism is the belief that nothing can be known for certain. The problem with this idea is that if nothing can be known for certain, then we could never be certain of the supposed truth of skepticism. Without a doubt, all this uncertainty is contrary to the simple doctrines of truth found in the restored gospel.

A better understanding of truth is the concept of absolute truth. Absolute truth is the notion that we can know things for certain. This concept is common in LDS theology. Brigham Young taught that truth "is based on eternal facts,"[12] and the scriptures declare that "truth is knowledge of things as they are, and as they were, and as they are to come" (D&C 93:24). "Things as they are" refers to facts about the world. When our beliefs correspond with the way things are or the facts of the world, we possess truth. When our beliefs do not correspond with the facts of the world, we possess falsehoods.

This straightforward understanding of truth is consistent with the teachings of the Lord. After describing truth as a "knowledge of things as they are, and as they were, and as they are to come" (D&C 93:24), He said "And whatsoever is more or less than this [understanding of truth] is the spirit of the wicked one who was a liar from the beginning" (D&C

93:25). In simple terms, truth is saying of what is, that it is, and saying of what is not, that it is not.

But what happens when two or more people claim to possess truth regarding the same situation, yet their claims are contradictory? What if one person says that the truth regarding a situation is "A" and another says that the truth regarding the same situation is "B"? Assuming that at least one of them is right, and that both of them cannot be correct, how do we know who has the truth? Critics of absolute truth argue that if there is no final arbiter of truth, then we cannot be certain of who is right and who is wrong; and if we cannot be certain, then absolute truth is impossible.

This argument against absolute truth is refuted by the fact that there is a final arbiter of truth. He is the omniscient Lord God of heaven and earth who sees and knows all through the Light of Christ. Because He is the fundamental source of truth and light, we can be certain that there is an absolute truth in everything, and that it exists in Him. Thus we see how important it is for us to understand that the Lord lives and that He is omniscient. People who do not appreciate this reality tend to fall prey to the "loosey-goosey" relativistic doctrines of truth.

The amount of secular knowledge we receive through the Spirit is commensurate with the amount of preparation, study, and sacrifice we put into an endeavor. In spiritual matters, where much is given, much is required (D&C 82:3); the same holds true in our search for secular knowledge. The blessings of inspiration usually come after we have demonstrated our faithfulness and dedication to an issue. Personal preparation and waiting on the blessings of the Spirit of the Lord may take a long time, but if we humble ourselves, keep His commandments, and petition Him for guidance, He will surely guide us toward the truth through the power of His Spirit.

Even after receiving knowledge through inspiration, there is always so much more for us to learn; our understanding is seldom complete. We must remember that the Lord gives knowledge "unto the children of men line upon line, precept upon precept, here a little and there a little" (2 Nephi 28:30). Thus, during times of uncertainty in our quest for secular truth, it may be prudent to talk about the verisimilitude of our ideas. Verisimilitude refers to the truth-likeness of our ideas, or the extent to which they represent reality. Ideas with high verisimilitude accurately represent reality, while those with low verisimilitude poorly represent reality. By adopting this perspective, we are acknowledging that there is always more to learn and that even our greatest ideas are continually evolving and are subject to change.

I am confident that if we persevere with a humble attitude, trust in the Lord, and do not reject the enlightening power of His Spirit, He will guide us to a greater understanding of the truth of all things. In order to achieve a greater verisimilitude in our secular ideas, then, we need to exercise humility, diligence, and trust in God.

FAITH AND REASON

The Lord has blessed mankind with two faculties for discerning truth: faith and reason. Faith and reason are interconnected. They may be thought of as two sides of the same coin. Faith requires reason, and reason requires faith. Paul's definition of faith illustrates their interconnectedness. He taught that "faith is the substance of things hoped for, the evidence of things not seen" (Hebrews 11:1). According to the second half of this definition, faith requires being able to reason that God lives given the evidence for His existence. In other words, faith requires being able to reason that, given the scriptures, prophets, spiritual experiences, and organization of nature, there is a Creator. When we witness this evidence, our powers of reason tell us that He lives.

Faith in God is not blindly accepting His existence without valid reasons. There is no such thing as "blind faith." Without evidence for the existence of a supreme being and the ability to reason that He lives, there is no faith; there is only ignorance. The Lord's teachings emphasize the important role that reason plays in developing faith. He said, "Let us reason together, that ye may understand; Let us reason even as a man reasoneth one with another face to face" (D&C 50:10–12). He has instructed those who preach His gospel to similarly reason with others so that they may also believe and develop faith in Him (D&C 66:7; 68:1).

Those who refuse to see the evidence for what it is and deny the Lord after receiving ample witness are sinning (Helaman 8:24). They have allowed themselves to become deceived by the craftiness of men that blinds their hearts and minds to the rational conclusion that there is a supreme being. Such was the case with Korihor, who lacked faith because he allowed himself to be deceived by the adversary. No matter how much Alma tried to reason with him—to help him "see" the evidence for what it is—he would not acknowledge God, that is, until he recognized the deception to which he had succumbed.

Though human reason is important for acquiring knowledge, it does not always lead to truth. Reason often requires faith in the Lord. Brigham Young said, "The whole world [is] under obligation to [God] for what they know and enjoy: they are indebted to him for it all."[13] Much of the knowledge that mankind receives is inspired by the Spirit of the Lord, and qualifying for inspiration requires faith. First Corinthians 2:14 teaches that the natural man, relying wholly on his own intuition and powers of reason, "receiveth not the things of the Spirit of God," because they are spiritually discerned. By itself, the ability to reason does not qualify us to receive the enlightening power of the Spirit of the Lord. We must believe in a higher power that is capable of inspiring our minds and directing us toward a greater truth.

The world would have us believe that reason is more certain than faith. This is not the case. The opposite is true. While reason may be faulty or sound, true faith is always sound because it is a belief in things "*which are true*" (Alma 32:21; emphasis added). Alma taught that when we have faith, our "knowledge is perfect in that thing" (Alma 32:34), and that our knowledge is real "because it is light" from the Spirit of the Lord (Alma 32:35). This certain knowledge is what enables us to testify that we *know* that God lives, notwithstanding having never seen Him. While evidence for the existence of God gives us reasons to believe in Him, it is our faith which gives us the certain knowledge that He lives.

Finally, Alma reminds us that although our faith can provide certain knowledge, our knowledge is not perfect (Alma 32:35–36). "Not perfect" means that our knowledge is not complete. There is always more to learn. Yet if we continually magnify our faith, our knowledge will grow toward a perfect understanding that reason alone is incapable of providing.

Notes

1. Galileo Galilei, *Discoveries and Opinions of Galileo*. Translated with an introduction and notes by Stillman Drake (New York: Doubleday Inc., 1957), 186.
2. *Encyclopedia of Mormonism*. Edited by Daniel H. Ludlow. Vols. 1–4. (New York: Macmillan, 1992), 1271.
3. Brigham Young, *Discourses of Brigham Young*. Selected and arranged by John A. Widtsoe (Salt Lake City: Deseret Book Co., 1954), 258.
4. Ibid., 391.
5. History of the Church, 2:396 (Feb. 17, 1836).
6. Ibid., p. 344 (Tues. Dec. 22, 1835).
7. *Encyclopedia of Mormonism*. Edited by Daniel H. Ludlow. Vols. 1–4 (New York: Macmillan, 1992), 1972.
8. Talk available at http://speeches.byu.edu.

9. George Q. Cannon, *Gospel Truth: Discourses and Writings of President George Q. Cannon.* Selected, arranged, and edited by Jerreld L. Newquist (Salt Lake City: Deseret Book, 1987), 284.
10. "Oh Say, What is Truth?" In the *Hymns of The Church of Jesus Christ of Latter-day Saints* (Corporation of the President of The Church of Jesus Christ of Latter-day Saints, 1985), 272.
11. Brigham Young, *Discourses of Brigham Young.* Selected and arranged by John A. Widtsoe (Salt Lake City: Deseret Book, 1954), 10:251.
12. Ibid., 14:115.
13. Ibid., 2.

TWO

DIVINE GUIDANCE THROUGH ANCIENT TIMES

Science and philosophy through all the ages have undergone change after change. Scarcely a century has passed but they have introduced new theories of science and philosophy, that supersede the old traditions and the old faith and the old doctrines entertained by philosophers and scientists. These things may undergo continuous changes, but the word of God is always true, is always right.

—Joseph F. Smith

Whenever we compare ancient and modern secular knowledge, there is, I think, a tendency to conclude that the ancients did not receive blessings of knowledge that are comparable to what we receive today. It is true that the outpouring of secular knowledge that has occurred over the last few hundred years is unmatched by other eras in the history of the world. Yet the scriptures teach us that Heavenly Father is no respecter of persons and that He loves all His children no matter when they live on the earth. Therefore, He must have provided the ancients with secular knowledge that enabled them to prosper and find joy. Historical evidence supports this claim. There is abundant evidence of the Lord blessing mankind as far back as the dawn of time. He provided the ancients with a great amount of secular knowledge essential to mankind's well-being.

In this section we'll review the secular knowledge that the ancients received from the Lord. We'll rely on scriptural and historical accounts to uncover the Lord's influence in one of the earliest known civilizations,

Sumeria, and look at the important role Abraham played in ancient Egypt's prosperity and knowledge. We will also discuss ancient Babylonian rulers who were called by the Lord to promote secular learning, and discover how their influence contributed to a new understanding of the world that led to the eventual overthrow of mythical worship.

SUMERIANS

The ancient land of Sumer was located above the Persian Gulf in the region of modern southwestern Iraq. Although a desert region today, in postdiluvian times it had an abundance of wetlands and fertile soil capable of producing vast amounts of grain. These ideal conditions played a major role in this region becoming the location for one of the earliest high civilizations. Sumerian civilization arose after the Flood, reached its height during the third millennium BC, and ended as a distinct society during the early second millennium BC.

Who were the Sumerians and where did they come from? According to the Jewish historian Flavius Josephus, the descendants of Shem, the Arphaxadites, settled the Mesopotamian region south of the Tower of Babel, known as Chaldea. Further evidence that Shem's descendants settled southern Mesopotamia is found in Sumerian cuneiform writing. The cuneiform writers used the word *Shumer*, not *Sumer*, to identify the region. Moreover, in accordance with the Sumerian phonetic law of admissibility, final consonants were not pronounced in speech. This means that the Sumerians most likely pronounced Shumer as "Shum." Also, the vowels "e" and "u" are interchanged in Sumerian and Hebrew languages, respectively. This means that the Sumerian word "Shum" becomes "Shem" when adopted into the Hebrew language.[1] It seems appropriate that the early Sumerians would name their land after their great forefather, Shem.

But what is the connection between the Sumerians and Hebrews, two civilizations separated by centuries and hundreds of miles? The answer is that the Father of the Hebrews, Abraham, came from Sumeria. Eight generations after Arphaxad, during the early second millennium BC, Abraham lived in Chaldea. He resided in the land of Ur, the place of his fathers' residence (Abraham 1:1). Abraham's Ur is generally accepted as the same region identified by archeologists as the ancient Sumerian city of Ur. These similarities enable us to confidently conclude that Chaldea and Sumer are the same region, and that Abraham's fathers, the Arphaxadite descendants of Shem, were early Sumerians.

We also know that the Sumerians' ancestors (the Arphaxadites) were a righteous people, for the descendants of the righteous Shem were entrusted with the Holy Priesthood, and Abraham recorded that his fathers were once an upright people. As we shall soon see, their righteousness and eventual apostasy provide an important context for understanding their secular accomplishments.

The archeological record (i.e., artifacts and cuneiform writings) clearly points out that the Sumerians had knowledge of metallurgy, textiles, pottery, and agriculture, all necessary to sustain a civilization. Linguistic analyses of cuneiform script and the names of Sumerian cities indicate that Sumerian knowledge of life-preserving skills originated with their predecessors.[2] Because of their proximity to the Flood (which appears in the Sumerian archeological record), we may deduce that these life-preservation skills came from antediluvian civilizations. And where did the antediluvian people get their knowledge from? Their knowledge came from the Lord.

Joseph Fielding Smith declared, "The first man was instructed by the best teacher man ever had, for he was taught of God. . . . It is true that he was left to work out, through the use of his faculties, many of nature's great secrets; but the Lord did not leave him helpless, but instructed him, and he was inspired by the Spirit of the Lord."[3] Joseph Smith similarly taught that "if there was anything great or good in the world, it came from God. The construction of the first vessel was given to Noah, by revelation . . . [and] the art of working in brass, silver, gold, and precious stones was taught by revelation, in the wilderness."[4]

Additional evidence suggesting that the Lord provided the Sumerians with secular knowledge comes from cuneiform texts. In their writings, the Sumerians acknowledged that a great deal of their knowledge came from deity. According to Sumerologist Samuel Noah Kramer, the earliest Sumerians believed in a supreme ruler of the heavens. This heavenly being, known to later Sumerians as Enlil, was deeply reverenced and held in high regard throughout Sumerian history. The Sumerians recognized him as "the king of heaven and earth"[5] and "the king of all the lands, the lord who determines ordinances."[6]

In addition, an ancient Sumerian hymn, "Enlil and the Creation of the Pickax," indicates that the Sumerians believed that Enlil was a benevolent deity who planned and created the "most productive features of the cosmos,"[7] "made the day come forth,"[8] "separated heaven from earth,"[9] "brought up 'the seed of the land' from the earth,"[10] "laid the plans which brought forth all seeds, plants, and trees from the earth,"[11] "established plenty, abundance, and prosperity in the land,"[12] and "fashioned the pickax and the plow as the prototypes of the agricultural implements to be used by man."[13] These people also believed that Enlil caused the Great Flood because of his displeasure with humanity.[14] Given his role as creator, judge, and provider of knowledge and prosperity, Enlil is most likely an apostate, late-Sumerian conceptualization of a being known to the early Sumerians (the Arphaxadites) as the true Lord of heaven and earth.

From the book of Abraham, we also know that the Lord provided these people with knowledge of the Creation (1:31). Abraham wrote that his righteous fathers received lessons on the Creation of the world, including the instruction that in the beginning, "the Spirit of the Gods was brooding upon the face of the waters" (4:2). Creation stories typifying this theme were likely passed on to successive generations, which explains the belief in water as the primal substance of the universe that once prevailed throughout ancient Mesopotamia.

We know that the Sumerians had a basic understanding of astronomy that enabled them to create a calendar that divided the year into summer (emesh) and winter (enten), and set the months by the lunar cycle. Some of their knowledge of the stars and planets also came from the Lord. Abraham states that his righteous fathers (the early Sumerians) received lessons on the stars and planets from the Lord (Abraham 1:31). Further, Abraham tells us that while he was in the land of Chaldea, he possessed scriptures on the Creation and astronomy (1:31), as well as the Urim and Thummim, from which he learned the nature and organization of the universe (3:1). Whether he shared this knowledge with his apostate neighbors is uncertain.

Moreover, the Sumerians were apparently aware of the "plants and roots" that the Lord has provided "to remove the cause of diseases" (Alma 46:40). Sumerologists discovered an ancient Sumerian medical tablet from the third millennium BC containing directions on how to make prescriptions from mineral, animal, and botanical sources. The tablet details how to combine natural ingredients to create medicines that were taken internally, and how to form pastes that were applied as a poultice to parts of the body.

What the early Sumerian medical tablet did not contain is equally revealing. Sumerologist Samuel Noah Kramer wrote that this "ancient [medical] document, it is worth noting, is entirely free from the magic spells and incantations which are a regular feature of the cuneiform medical texts of later days." He continued, "The physician who wrote this document, therefore, seems to have practiced his medicine along empirico-rational lines."[15] The absence of mystical practices on the third millennium BC medical tablet suggests that the earlier Sumerians were not given to idol worship.

The notion that early Sumerians were not given to idol worship is corroborated by Abraham's assertion that his ancestors were once a righteous people. He points out that his fathers once followed the "holy commandments which the Lord their God had given unto them" (1:5). The claim that many of the early Sumerians were righteous is also corroborated by the archeological record. Kramer wrote that, "according to their own records," these people "cherished goodness and truth, law and order, justice and freedom, righteousness and straightforwardness, mercy and compassion, and naturally abhorred their opposites, evil and falsehood, lawlessness and disorder, injustice and oppression, sinfulness and perversity, cruelty and pitilessness."[16]

In sum, the ancient Sumerians were indeed blessed with secular knowledge from the Lord. He provided their antediluvian ancestors with knowledge of natural remedies, medicine, agriculture, pottery, and smelting, and this knowledge was passed down to them. Their archeological record indicates that they recognized deity as the main source of life and knowledge. The Lord gave the early Sumerians (Abraham's fathers) principles of righteousness which enabled them to live in relative peace and harmony and to lay the foundation of a great civilization.

Unfortunately the people eventually fell away from righteousness. The seeds of apostasy were probably sown a few centuries before Abraham, most likely during the early to middle third millennium BC when Sumerian society became increasingly polytheistic.[17] By the time Abraham arrived on the scene during the late third to early second millennia BC, the spiritual apostasy was in full force. Many people had turned away from righteousness, rejected the commandments of the Lord, and were worshipping heathen gods (Abraham 1:5).

As the Sumerians fell into apostasy, their blessings of peace and prosperity were taken away. During the late third to early second millennia Third Dynasty of Ur, privation and starvation afflicted the people. The Sumerian record indicates that the city of Ur was afflicted by famine during this era,[18] as does the book of Abraham, which tells of a famine in the land of Ur (Abraham 2:1). Enemy incursions were also troublesome during the Third Dynasty of Ur. Sumeria was beleaguered by invasions from the Semitic peoples to the north and the Gutian barbarians from the mountains to the east.

During one invasion, Elamites from the mountainous region east of Babylonia (see Bible Dictionary), took the Third Dynasty king, Ibbi-Sin, into captivity and destroyed the temple of Ur. A Sumerian poet recorded the great lamentation that took place after the temple's destruction.[19] From the poet's account, it is apparent that the Sumerians considered the destruction of their mythical religious symbols to be grievous events.[20] This claim is corroborated by the book of Abraham, which records that after the "Lord broke down the altar of Elkenah, and of the gods of the land, and utterly destroyed them, . . . there was great mourning in Chaldea" (Abraham 1:20).

In all likelihood, the culminating event in the Apostasy was the conquest of Sumeria by the Babylonian king, Hammurabi. Hammurabi's conquest of Chaldea effectively brought Sumerian civilization to an end in 1750 BC. Abraham escaped the wars, famines, and wickedness that plagued Sumeria by leaving the land of Chaldea and journeying to Egypt where, according to the Lord's instructions, he expanded the Egyptians' understanding of numbers and the heavens.

EGYPTIANS

According to the writings of Josephus and Abraham, the first postdiluvian settlers in the Nile region descended from Ham. Josephus wrote that the Egyptians were the descendants of Mesraim, the son of Ham,[21] and Abraham wrote that the first government in Egypt was established by Pharaoh, the eldest son of Egyptus, herself being a daughter of Ham who came from the Ark (Abraham 1:21–23). Unlike the people of Shem in Chaldea/Sumeria, the people of Ham did not have the rights of the priesthood (Abraham 1:27); nevertheless, the Lord blessed them with secular knowledge that made possible marvelous achievements.

Through the writings of Abraham, we know that Pharaoh, the eldest son of Egyptus, ruled in righteousness and sought "earnestly to imitate that order established by the fathers in the first generations . . . even in the reign of Adam, and also of Noah" (Abraham 1:25–26). If the generational proximity of the early Egyptians to Adamic civilizations enabled them to retain the first order of government established by man, then certainly they would have retained an understanding of smelting, tool making, agriculture, construction, and time reckoning as well. This suggests that many of the life-preserving skills that the ancient Egyptians possessed came from their antediluvian predecessors who, as we have seen, received their knowledge directly from the Lord.

Ancient Egypt's knowledge made possible fantastic agricultural and engineering achievements, such as an ambitious building program which led to the construction of awe-inspiring pyramids. One of these pyramids, the Great Pyramid of Giza, is the only wonder of the ancient world still standing today. It is truly an engineering marvel. Consider that each of its four sides is approximately 756 feet in length, and that taken together, the four sides make a near perfect square. The difference between the longest and shortest sides is a mere 7.9 inches! Moreover, because a natural outcrop of rock was allowed to remain near the center of the base, the builders had to measure the location of the sides along the outside rather than through the middle, which made aligning the base a difficult task.

What makes the construction of the pyramid more astounding is that the ascending outside edges appear straight, and the four sides are nearly perfectly oriented to the four polar points, north, south, east, and west. In fact, the alignment of the four sides deviates from the true cardinal points by only fractions of a degree. Because the compass had not been invented yet, the builders had to align the sides by the stars. Achieving this degree of precision on such a large structure would be a magnificent engineering achievement even by today's standards.

Clearly the builders possessed skills in tool making and stone masonry. As has been shown, their understanding of these skills came from their antediluvian ancestors, who received it from the Lord. But where did their knowledge of mathematics and astronomy that was necessary for precise measurement and orientation come from? We know from a few sources that this knowledge was not possessed by the ancient Egyptians before Abraham entered the Nile region.

Abraham wrote that after he left his homeland in Chaldea, he made his way to Egypt, where he was instructed by the Lord to share his knowledge of the stars, planets, and reckoning of time with the Egyptians (Abraham 3:15). Josephus similarly wrote that while Abraham was in Egypt, "he communicated to them arithmetic, and delivered to them the science of astronomy; for, before Abram came into Egypt, they were unacquainted with those parts of learning; for that science came from the Chaldeans into Egypt."[22] These accounts are corroborated by the prophet Joseph Smith's declaration that "The learning of the Egyptians, *and* their knowledge of astronomy was no doubt taught them by Abraham and Joseph . . . who received it from the Lord."[23]

From these sources we may conclude that if the Egyptians built the Great Pyramid, their knowledge of mathematics and astronomy that was used in its construction came from Abraham. However, placing the construction of the pyramid after the arrival of Abraham into Egypt contradicts the commonly accepted timeline for the pyramid's completion. Egyptologists claim that the pyramid was completed during the mid third millennium BC (circa 2560 BC); however, Abraham lived during the early second millennium BC. If the pyramid was completed in 2560 BC, then the Egyptians would have had a good grasp of astronomy and mathematics centuries before Abraham was born.

If the pyramid was built without the benefit of Abraham's teachings, where did the knowledge of mathematics and astronomy necessary for its construction come from? There is evidence to suggest that this knowledge came from foreigners who once lived in the Nile region. The fifth century BC historian Herodotus wrote that the Egyptians called "the pyramids after Philition, a shepherd who at that time fed his flocks about the place" where the pyramid stands, during the reign of Cheops.[24] During that time the rulers over Egypt "shut up and never opened" the Egyptian temples, which were symbols of idolatrous worship. Herodotus wrote that the "Egyptians so detest[ed] the memory of these kings that they [did] not much like even to mention their names."[25]

Who were these kings that closed the temples around the time that the great pyramid was constructed? Writings from the third century BC Egyptian historian Manetho provide some clues.

> There was a king of ours, whose name was Timaus. Under him it came to pass, I know not how, that God was averse to us and there came, after a surprising manner, men of ignoble birth out of the eastern parts, and had boldness enough to make an expedition into our country, and with ease subdued it by force, yet without our hazarding a battle with them. So when they had gotten those that governed us under their power, they . . . demolished the temples of the gods.[26]

Manetho called these foreigners "Hycsos" or "shepherd-kings." The shepherd-kings ruled for several centuries until they were driven out by an Egyptian king, at which time they headed north toward Syria. Manetho recorded that as they journeyed northward they "were in fear of the Assyrians, who had then the dominion over Asia." Therefore, "they built a city in that country which is now called Judea," a city large enough to be defended by a "great number of men, and called it Jerusalem."[27]

It is important to note that knowledge of mathematics and astronomy was not readily available to the ancients during the third millennium BC. Mathematical and astronomical understanding was largely limited to people who received it through inspiration. If the shepherd-kings were the architects of the pyramid, then they must have acquired their understanding of mathematics and astronomy from the Lord. Evidence that the shepherd-kings were a righteous people who may have received this knowledge through inspiration comes from Manetho's claim that the shepherd-kings attempted to curb idolatry by closing Egypt's temples.[28]

Who were these shepherd-kings and where did they come from? No one knows for sure, though we may speculate on their identity. They could have been a righteous group of Arphaxadites from the land of Chaldea/Sumer. We know that Abraham's ancestors received knowledge from the Lord, and archeological evidence of calendrical rendering and numerical tables indicate that these people understood mathematics and astronomy. Or perhaps the shepherd-kings were associated with Job, a man described in the Bible as a righteous shepherd-king from the east. If Job was involved in the construction of the Great Pyramid, it might explain why he was rebuked by the Lord for becoming too prideful in his accomplishments (Job 38:1–7). On the other hand, it may be that the shepherd-kings were associated with Melchizedek and his fathers. Melchizedek was a man of exceeding faith who ruled over Salem (Jerusalem) during a time when the Assyrians dominated Asia. These details correspond with Manetho's assertion that the shepherd-kings founded Jerusalem while the Assyrians ruled Asia. Finally, Josephus believed that the shepherd-kings were the children of Israel. Identifying the shepherd-kings as the House of Israel is problematic because the pyramid is believed to have been built centuries before Jacob and his people arrived in Egypt; nevertheless, this theory is not without merit. The children of Israel kept flocks and herds (Genesis 50:8) in the land of Egypt, Joseph's status as second-in-command qualified him as a ruler, and the 430 years that the children of Israel dwelt in Egypt (Exodus 12:40) approximates Manetho's estimation that the shepherd-kings stayed in the land for five centuries.[29]

We may never know the identity of the pyramid builders and shepherd-kings. One thing we can be certain of, however, is that when Abraham arrived in Egypt, he encountered a civilization with a limited understanding of mathematics and the heavens. He shared his knowledge of these subjects with these Egyptians according to the will of the Lord, illustrating once again that the Lord provided the ancients with valuable secular knowledge.

MESOPOTAMIA

The Lord's blessings to the ancient Sumerians and Egyptians were not limited to secular knowledge; He also blessed them with spiritual knowledge. In the book of Abraham we read that these ancients possessed principles of righteousness. Abraham wrote that his fathers were righteous (1:5) and that the first pharaoh of Egypt ruled in righteousness (1:26).ABraham must have learned about their earlier righteousness through revelation or historical records because, when he arrived on the scene in Sumeria and Egypt, the people had forgotten the Lord and turned to idolatry (1:27).

As Sumerian and Egyptian idolatry spread throughout Mesopotamia and the surrounding region, people increasingly turned to mythological beliefs and practices. Even the House of Israel was not immune from this influence. Following their Exodus from Egypt, many built a golden calf and wished to return to idolatrous Egypt, and after cleansing the promised land of Canaanite mythological influences, the children of Israel had to repeatedly repent for idol worship.

Because of their idolatry, many ancient people interpreted the natural world in terms of a pantheon of gods and believed that natural events were influenced by the whims of the gods of the sky, earth, and water, to name a few. Their focus on understanding and appeasing the will of the gods meant that a study of the natural world with the intent of discovering the laws of nature and its Lawmaker probably never entered the minds of many. Thus idolatry prevented the ancients from learning about the Lord and the natural order of His creations.

The ancient world floundered in this state of spiritual and intellectual darkness for a long time, but the Lord did not allow it to continue indefinitely. He gradually brought the ancient world out of darkness by enlightening scholars' minds and raising up rulers who promoted secular discovery. This change was evident in ancient Babylon.

BABYLONIANS

Notwithstanding the idolatry that was rampant in their society, Babylonian scholars made progress toward understanding nature as ordered. They made strides in astronomy and mathematics, as evidenced by having developed a calendar based on the sun's movement and the phases of the moon. They even used mathematical procedures to predict eclipses—an impressive achievement even by modern standards. It is important to note, however, that the scholars were not wholly responsible for this progress; a couple of noteworthy Babylonian rulers encouraged secular learning.

Hammurabi was one of the most influential rulers in Babylon. His rule over Babylon lasted some forty years, from approximately 1792 BC to 1750 BC. One of Hammurabi's greatest contributions was the secular code of laws he established to govern Babylonian society. The code contained 282 provisions dealing with issues such as labor, property ownership, trade and commerce, and family. In many respects, his code represented the law of Moses. John A. Widtsoe described the code as "contain[ing] injunctions for correct living resembling the Ten Commandments."[30] He attributed Hammurabi's virtuous injunctions to principles of righteousness that had been taught and passed down since the dawn of time, notwithstanding apostasy from spiritual truth.[31] For example, under family law are provisions stating that those who wish to marry must enter into a written contract and that a divorced father should pay alimony and child support. There are also provisions on crime and punishment, such as that legal cases are to be initiated in writing, testimonies are to be taken under oath, witnesses may be subpoenaed, and punishment should fit the crime.

Hammurabi created the laws to bring stability and security to the people of Mesopotamia. In 1902 archeologists discovered a pillar containing the code and an inscription from Hammurabi himself. Hammurabi inscribed that his laws were intended to "bring about the rule of righteousness in the land, to destroy the wicked and the evil-doers; so that the strong should not harm the weak; . . . [to] enlighten the land, [and] to further the well-being of mankind."[32] Without a doubt, Hammurabi's laws ensured that people's lives were not wholly driven by the demands of depraved priests and the whimsical "wishes" of false idols. His laws also provided the people with a degree of stability that enabled some to spend more time learning than fighting for survival.

The thousand years following the reign of Hammurabi were tumultuous times for Mesopotamia. Babylon was dominated by foreign rulers, many who were from the Assyrian Empire to the north. Then in 612 BC, a Chaldean named Nabopolassar seized control of Babylon from the Assyrians. After ascending to power, Nabopolassar implemented an ambitious temple-building project in an attempt to placate the citizens of Babylon. One of his projects included the rebuilding of the great ziggurat called Etemenanki, believed to be the biblical tower of Babel.

While the exact purpose of the Etemenanki and similar smaller ziggurats is unknown, many historians believe that the Mesopotamia priests used these structures to get closer to heaven so that they could be better heard by the gods. To show his respect for local customs, Nabopolassar had some of his sons haul bricks for the rebuilding of Etemenanki.[33] One son who may have helped in this effort was Nebuchadnezzar.

In 605 BC, Nebuchadnezzar succeeded his father and became king of Babylon. According to secular and biblical history, Babylonia became even more powerful under the reign of Nebuchadnezzar. Nebuchadnezzar continued the ambitious building projects his father started and constructed one of the wonders of the ancient world, the fabled Hanging Gardens.

Nebuchadnezzar is also known to have encouraged learning and discovery. His interest in scholarly work is evidenced by his having commissioned the children of Israel who possessed great wisdom and knowledge to serve him after the house of Judah was taken into captivity (circa 587 BC). One of those taken captive Babylon was Daniel. When the king's master of the eunuchs, Ashpenaz, recognized Daniel as "skillful in all wisdom, and cunning in knowledge and understanding science," he and his companions were taught the language of the Chaldeans so that they would be able to share their wisdom with the king (Daniel 1:4). These events clearly show that Nebuchadnezzar was concerned about acquiring knowledge.

There is also good reason to believe that the scholarly-minded Nebuchadnezzar distrusted mythology, as evidenced by his actions after having a dream of a stone that was cut without hands. Nebuchadnezzar was so troubled by this dream that he commissioned the wise men (magicians, astrologers, and sorcerers) in his kingdom to tell him the dream and give an interpretation. When the wise men asked the king to describe the dream, he replied that he could not remember the details. I believe that the king feigned not being able to remember the dream because he was doubtful of the magicians', astrologers', and sorcerers' claims to truth. In other words, he was testing them.

When the wise men replied that no man could provide an interpretation without hearing the details of the dream, the king was "angry and very furious" and commanded that they be put to death (Daniel 2:12). Some might think that Nebuchadnezzar went too far by ordering their deaths. I do not think that this is the case. He was a scholarly-minded king, and as such, he valued knowledge. I believe that he was suspicious of their claims to truth; he had grown tired of these men wandering throughout his kingdom professing access to knowledge by means of divination. At last he had exposed their ruse; the gig was up. They were caught "prepar[ing] lying and corrupt words to speak before [him]" (Daniel 2:9). The punishment for intentionally deceiving the king was death.

Daniel was granted permission to speak to the king after the decree had been given. Perhaps someone reminded the king of the Hebrew wise men and convinced Nebuchadnezzar to let Daniel speak. Or perhaps the king was impressed with the fact that one of his wise men was brave enough to approach him, while other wise men were leaving the kingdom or suddenly changing careers. In any event, Daniel told Nebuchadnezzar that the dream was of a personage with a head of gold, breast and arms of silver, belly and thighs of brass, legs of iron, and feet of both iron and clay. He explained that the head of gold represented Nebuchadnezzar's kingdom, that the parts of silver, brass, iron, and iron and clay represented kingdoms that would follow, and that the great stone cut out of the mountain without hands represented the gospel that would fill the earth in the latter days.

Not surprisingly, Nebuchadnezzar's influence created a "golden age" for Babylonia. Advances in astronomy and mathematics that occurred in Mesopotamia during and after his reign were a direct result of his desire for truth and disdain for mystical explanations. Toward the end of his reign, Nebuchadnezzar came to the realization that he alone was not responsible for this progress. He acknowledged the divine influence of the Lord when he declared, "I blessed the most High, and I praised and honoured him that liveth forever, whose dominion is an everlasting dominion, and his kingdom is from generation to generation" (Daniel 4:34).

The prophet Daniel also acknowledged the Lord's role in calling up rulers and giving knowledge to mankind when he exclaimed, "Blessed be the name of God forever and ever: for wisdom and might are his: And he changeth the times and seasons: he removeth kings and setteth up kings: he giveth wisdom unto the wise, and knowledge to them that know understanding" (Daniel 2: 20–21).

I believe that the Lord called the scholarly-minded Hammurabi and Nebuchadnezzar to prepare the way for a great secular illumination that would eventually lead to the overthrow of idolatry and prepare the inhabitants of the earth for the gospel of Jesus Christ.

Notes

1. A Poebel, *American Journal of Semitic Languages*, LVIII, 20–26.
2. Samuel Noah Kramer, *The Sumerians: Their History, Culture, and Character* (Chicago: University of Chicago Press, 1963), 41.
3. Joseph Fielding Smith, *Doctrines of Salvation*. Edited by Bruce R. McConkie (Salt Lake City: Bookcraft, 1954–56), 1:95.
4. Joseph Smith, *Discourses of the Prophet Joseph Smith*. Compiled by Alma P. Burton (Salt Lake City: Deseret Book Co., 1977), 201.
5. Samuel Noah Kramer, *The Sumerians: Their History, Culture, and Character* (Chicago: University of Chicago Press, 1963), 119.
6. Ibid., 336.
7. Ibid., 119.
8. Ibid.
9. Ibid., 145.
10. Ibid.
11. Ibid., 119.
12. Ibid.
13. Ibid.
14. Alexis Q. Castor, *Between the Rivers: The History of Ancient Mesopotamia* (Teaching Company, 2006), 28.
15. Ibid., 97.
16. Ibid., 123.
17. Ibid., 118, 141.
18. Ibid., 70.
19. Ibid., 142.
20. Ibid.
21. Flavius Josephus, *Josephus: The Complete Works*. Translated by William Whiston, A.M. (Nashville: Thomas Nelson, 1998), 1.6.2.
22. Ibid., 1.8.2.
23. Joseph Smith, *Teachings of the Prophet Joseph Smith*. Selected and arranged by Joseph Fielding Smith (Salt Lake City: Deseret Book, 1976), 251; emphasis added.
24. Herodotus, History, Euterpe (Translation by George Rawlinson).
25. Ibid.
26. Flavius Josephus, *Against Apion*. In *Josephus: The Complete Works*. Translated by William Whiston, A.M. (Nashville, TN: Thomas Nelson, 1998), 1.14.
27. Ibid.
28. Ibid.
29. Ibid.
30. John A. Widtsoe, *Evidences and Reconciliations* (Salt Lake City: Improvement Era, 1960), 34.
31. Ibid., 37.
32. Code of Hammurabi.
33. L. Sprague De Camp, *The Ancient Engineers: Technology and Invention from the Earliest Times to the Renaissance* (Barnes & Noble Books, 1993), 70.

THREE
THE FIRST GREAT ILLUMINATION

Their souls were illuminated by the light of the everlasting word.
—Alma 5:7

THE GOLDEN AGE OF GREECE

Pre-Socratics

During the early sixth century BC, as the Babylonians were being conquered by Cyrus's Persian Empire, the Lord was preparing a small group of people in a rugged land on the borders of the Aegean Sea for an intellectual illumination that would change the world. The secular illumination that took place among the Greeks led to the decline of idolatry, brought the world out of intellectual darkness, and prepared the gentiles for the gospel of Jesus Christ. The following discussion demonstrates that the Lord's Spirit was moving among these people and inspiring their scholars to bring to pass the purposes of God.

Around 600 BC, when Lehi and his family were leaving Jerusalem, the Spirit of the Lord inspired a man named Thales to leave the Greek-Ionian merchant town of Miletus and embark on a journey for truth. Thales traveled to Mesopotamia and Egypt, where he studied the astronomy and mathematics that had been given to those people by the Lord, and then took the things he learned back to his homeland. Thus we see how, as Josephus claimed, "science came from the Chaldeans into Egypt, and from thence to the Greeks also."[1]

Back in Miletus, Thales set up a school that taught measurement, engineering, and astronomy, and emphasized that the natural world is ordered and predictable. The notion of an ordered world conflicted with the belief in a world influenced by the whims of mythical gods, such as that a violent sea manifested Poseidon's displeasure and that Zeus' anger caused thunder and lightning. This change in thinking at Miletus was a decisive moment in the intellectual history of the world; it marked the beginning of an intellectual revolution that eventually changed the way mankind viewed the natural world and its processes. It also pointed humanity in the direction of better understanding the divinely appointed laws of nature.

Thales studied the fundamental nature of the universe and the principles on which it operates, an endeavor known as natural philosophy. Natural philosophy represented a fundamental shift in thinking about the universe. Back then, scholars concerned with cosmology speculated on the order of the world and the heavens, something that they knew little about because they lacked understanding. The introduction of Thales's natural philosophy was a monumental event because it shifted scholars' focus from unknowable cosmological speculations to something they could know through observation, the physical nature of the world.

Thales's theory on the fundamental nature of the physical world (physis) was that water is the primal substance. Centuries later, Aristotle speculated that Thales believed that water was the primal substance because of its obvious importance in sustaining life. It is also possible that Thales's physis was influenced by Mesopotamian accounts of creation which identify water as a primal substance, accounts which likely originated with the early Sumerians who received lessons on the Creation from the Lord before falling into apostasy.

During the sixth century BC, Greek scholars such as Anaximander, Anaximenes, Heraclitus, Parmenides, Empedocles, and Antaxagoras continued Thales's work. These pre-Socratics, as they are often called, relied on observation, rational discussion, and reason in their quest for knowledge about the nature of the world. Their approach was revolutionary because it did not resort to mythological and supernatural explanations. It helped popularize the belief that people could rely on reason and observation without resorting to priestcraft. There was no place for mythological gods in pre-Socratic natural philosophy. Natural philosophy first gained approval among educated Greeks who no longer accepted mythical explanations of natural events. But for others, change would take much longer. When the Apostle Paul arrived in Athens Greece six centuries later, he found the city still "wholly given to idolatry" and full of superstition (Acts 17:16, 22).

The pre-Socratics' rejection of mythical explanations left a void in Greek scholarship. If the natural world was not driven by the whims of mythical gods, then who or what was controlling the world? Two theories emerged in pre-Socratic philosophy. First, scholars like Leucippus and Democtritus adopted an atheistic doctrine known as atomism. Greek atomism was the belief that there was no supreme creator, only invisible and indivisible particles of matter. They believed that these particles, which they called atoms, combined to form physical reality. They also believed that because the form and flux of atoms composed ultimate reality, there was no divine influence or purpose in the universe; there were only atoms. The principles of atomism became the "higher power" for some pre-Socratics, much in the same way that many modern atheists revere the physical laws of nature without acknowledging a divine author.

On the other hand, there were pre-Socratic scholars like Anaxagoras and Xenophanes who filled the void left by mythology with monotheism. Xenophanes (circa 500 BC) introduced monotheism into pre-Socratic philosophy when he claimed that there is a divine force at work in the universe, and Anaxagoras claimed that a divine mind (called Nous) brought direction, life, and purpose to the cosmos. This introduction of monotheistic cosmologies in mainstream pre-Socratic scholarship is truly amazing! At the time, Greek civilization was not familiar with monotheistic theologies. Of the two dominant monotheistic religions that existed during that time, Judaism was largely confined to Palestine on the eastern shores of the Mediterranean, and Zoroastrianism had not yet spread beyond the borders of Persia.[2] Moreover, Christianity would not arrive in Greece for another six centuries when the Apostle Paul visited that region.

Why did pre-Socratics such as Xenophanes and Anaxagoras choose to accept a supreme creator? Without a doubt, they were inspired by the Spirit of the Lord. Daniel 2:21 says that the Lord gives "knowledge to them that know understanding." The Lord blessed the pre-Socratics with an appreciation of a supreme being because they "knew understanding." They knew that mythological gods played no role in nature and, most important, they knew the value of truth and desired to obtain it. Greek scholar Charles Freeman described the intellectual integrity of the pre-Socratics as follows:

What united them was their readiness to challenge existing conventions, to dig down to bedrock in their attempt to discover what may be called the truth. They raised questions about the underlying purpose and operation of the physical world and the nature of reality. They began to explore the tools with which truth could be found, language, the senses, the use of reason, and see their uses and inadequacies. These remain the central concerns of philosophy to this day. It was a remarkable achievement.[3]

To be wholly dedicated to revealing truth requires humility—humility to accept that your preconceived notions may be wrong, and to accept discoveries contrary to your expectations. As pre-Socratics like Xenophanes and Anaxagoras humbled themselves in the pursuit of truth, I believe they came to the realization that there is a higher power or supreme force at work in the universe. In their search for truth about nature, they most likely experienced that "all things denote that there is a God; yea even the earth, and all things that are upon the face of it, yea, and its motion, yea, and also the planets which move in their regular form do witness that there is a supreme creator" (Alma 30:44). The Spirit of the Lord was the source of this knowledge. The inspiration it provided was part of God's plan to enlighten mankind both spiritually and secularly.

SOCRATES, PLATO, AND ARISTOTLE

So great was Socrates's influence on Greek scholarship that we refer to the Greek scholars who preceded him as pre-Socratics. Socrates continued the intellectual tradition of his pre-Socratic predecessors by rejecting mythical explanations in the search for truth. However, rather than focusing on the natural world as they had done, he focused on the essence of moral concepts such as truth, justice, virtue, beauty, and how to live a meaningful life.

Socrates's most important contribution was emphasizing reason as a pathway to truth. He encouraged others to embrace rational skepticism, a hallmark of scientific thought, and reject truth claims by idolatrous priests. Unfortunately, his efforts to get others to think rationally and question the mythological status quo got him into trouble with the rulers of Athens. He was indicted for "rejecting the gods acknowledged by the state" and for "corrupting the youth" with his teachings.[4] He was found guilty of the aforementioned charges in what was most likely a politically charged trial, and sentenced to death by drinking hemlock, a poison which paralyzes the motor nerves and causes asphyxiation by disrupting the diaphragm.

Socrates spent a month in prison before his death sentence was carried out. During that time he regularly visited with friends, students, and family. Prison security was so relaxed that Socrates's lifelong friend, Criton, encouraged him to escape. According to science historian George Sarton, the Athenian judges may have welcomed his escape, but Socrates would not leave. He believed that "injustice [could] not be corrected with injustice" and that it was a citizen's duty to obey the laws of the city even if those laws were unjust.[5]

While in prison, he continued to conduct himself in a dignified manner, so much so that he won the admiration of his prison attendant. In the *Phaidon*, Plato wrote that when the attendant arrived to tell Socrates that it was time to take the poison, he said:

> Socrates, I shall not find fault with you, as I do with others, for being angry and cursing me, when at the behest of the authorities, I tell them to drink the poison. No, I have found you in all this time in every way the noblest and gentlest and best man who has ever come here, and now I know your anger is directed against others, not against me, for you know who are to blame. Now, for you know the message I came to bring you, farewell and try to bear what you must as easily as you can.[6]

After speaking these words, the attendant burst into tears and started out of the room. As he left, Socrates replied, "Fare you well, too; I will do as you say." Then he turned to his friends and said, "How charming the man is! Ever since I have been here he has been coming to see me and talking with me from time to time, and has been the best of men, and now how nobly he weeps for me!"[7]

Why did Socrates conduct himself in such a noble manner given the circumstances leading up to his death? The answer is that he had the conviction of knowing that he was following the will of a higher power. He often stated that he was being guided by a divine voice. Consider, for example, this comment on his work and subsequent conviction. He stated:

> Know that the God commands me to do this, and I believe that no greater good ever came to pass in the city than my service to the God. For I go about doing nothing else than urging you, young and old, not to care for your persons or your property more than for the perfection of your souls, or even so much; and I tell you that virtue does not come from money, but from virtue come money and all other good things to man, both to the individual and to the state. If by saying these things I corrupt the youth, these things must be injurious; but if anyone asserts that I say other things than these, he says what is untrue. Therefore I say to you, men of Athens, ... that I shall not change my conduct even if I am to die many times over.[8]

Some of his teachings alluded to in this quote bear a striking resemblance to restored gospel doctrine. He believed that self-knowledge is essential for acquiring personal virtue, and that ignorance leads to evil. This belief is consistent with latter-day scripture, which instructs us to "seek learning" (D&C 88:118) so that we may grow and become more like Him whose glory is intelligence (93:36). Also, his belief that it is more important to care for the perfection of our souls than our physical possessions imitates the Lord's warning against acquiring physical possessions at the risk of losing our souls (Mark 8:36). That Socrates's moral system "has not been cancelled or diminished by the progress of Christianity"[9] is a testament to his inspired beliefs.

Socrates's emphasis on rational argument and logical discovery not only laid the groundwork for future scientific endeavors, but it also strengthened the cause of truth by turning mankind away from mythical beliefs and practices. As we shall see in the next chapter, his attempts to promote rational thinking helped prepare the Greeks to receive the gospel of Jesus Christ that would be taught throughout their land centuries later.

Socrates's student, Plato, similarly embraced a rational quest for truth. Plato's philosophical contributions helped popularize an investigation of ontological issues concerning what is real in the world, and epistemological issues concerning how we acquire truth about the world. According to his theory of Forms, he believed that earthly objects and concepts have perfect counterparts that exist in a metaphysical domain called the world of Forms, and that truth regarding the world of Forms could be discovered through reason. An equally important contribution was establishing a center of learning in Athens. At his academy in Athens, Plato fostered both a love of learning and intellectual freedom. He encouraged his students to explore their own ideas and avoid blind acceptance of someone else's belief system, including his own. Out of this positive learning environment came the culminating figure of the first illumination, Aristotle.

Aristotle (384–322 BC) joined Plato's Athens Academy at age seventeen. He stayed at the Academy for twenty years until Plato's death in 347 BC. Aristotle believed that there was an absolute truth about everything in the world. He adopted a viewpoint on the fundamental essence of the physical world that was similar to Plato's, but differed with Plato on how to discover that essence. Whereas Plato believed that truth could largely be discovered through reason, Aristotle claimed that, because the mind could only reason on that which had been experienced by the senses, both reason and observation were equally important for revealing truth about the natural world.

It has been said that Aristotle was the last person who knew everything there was to know, during his lifetime. This may be true. His writings that have survived demonstrate a vast knowledge of a wide variety of topics such as politics, poetics, rhetoric, ethics, logic, physics, and biology. Equally impressive were his theological and cosmological ideas. He believed in a supreme creator, or "unmoved mover," who set nature in motion and is the cause of all things, and that nature follows a grand purpose.

Aristotle also believed that each entity in nature has a purpose, which he called entelechy, and that all living things have a life-giving soul which determines their standing in the hierarchy of nature. For example, he thought that the soul of a plant differs from that of an animal, and that both differ from the soul of a human, who is closer to the unmoved mover. The closer an entity is to the unmoved mover, the more perfect its state.

Some of Aristotle's views resemble gospel teachings. He believed in a supreme being that created the natural world for a divine purpose, and his notion of a hierarchy among living things is similar to the gospel doctrine that there are "classes of beings in their destined order or sphere of creation" (D&C 77:3). His belief that the human soul is more similar in kind to the unmoved mover than other living things, yet lower than the unmoved mover, is consistent with LDS doctrine which asserts that some spirits are more intelligent than others and that the Spirit of the Lord is "more intelligent than they all" (Abraham 3:19).

The fact that Aristotle developed these viewpoints within a society steeped in superstition and mythology is remarkable. Why did he choose to accept a monotheistic supreme creator while many of his countrymen continued to worship dumb idols? Certainly he was influenced by his monotheistic predecessors, but Aristotle was an independent thinker, and as such, he would not have accepted monotheism if he did not believe it. Most likely, his convictions about an unmoved mover came from his passion for revealing truth and his extensive study of nature. Consider these words from the book of Job.

> But ask now the beasts, and they shall teach thee, and the fowls of the air, and they shall tell thee: Or speak to the earth, and it shall teach thee: and the fishes of the sea shall declare unto thee. *Who knoweth not in all these that the hand of the Lord hath wrought this?* In whose hand is the soul of every living thing, and the breath of all mankind. (Job 12:7–10; emphasis added)

Indeed, as Aristotle carefully studied nature in the pursuit of truth, the Spirit of the Lord whispered to his mind that a supreme being created, organized, and sustains the world.

Overall, the intellectual accomplishments of Greek scholars from the pre-Socratics to Aristotle were so remarkable that this time period is often referred to as the Golden Age of Greece. These scholars' accomplishments put mankind on the path to recognizing that the natural world is ordered and regular, and led to the eventual overthrow of mythical interpretations of natural processes. This change paved the way for humanity to more fully recognize the Lord as the creator and to discover that He "hath given a law unto all things [in nature], by which they move in their times and seasons" (D&C 88:42). Philosopher of science Karl Popper aptly commented that "it is almost too good to be true [that in] . . . every generation [during the Golden Age] we find at least one new philosophy, one new cosmology of staggering originality and depth."[10] Indeed, these accomplishments comprise nothing less than a Greek miracle.

Why did this intellectual miracle take place in Greece and not somewhere else? There are no definitive answers to this question. While the presence of democratic institutions and emphasis on critical and open debate certainly played a role, these do not fully explain why the revolution occurred in Greece and not somewhere else. All we can be certain of is that the intellectual revolution occurred in ancient Greece because that is where the Lord chose to pour out His Spirit. His Spirit inspired Greek scholars to search for truth through reason and observation, rather than mythology. As we shall soon see, this enlightening power prepared the Greek gentiles for the gospel of Jesus Christ.

GREEK RULERS

The first illumination that occurred during the Golden Age of Greece resulted in a boom of rational and empirical discovery among Greek scholars that lasted from 300 BC to AD 200. The Lord's influence on secular scholarship during this five-hundred-year period is especially evident in the Greek city of Alexandria, in Egypt, during the rule of the Ptolemaic kings. But before examining the Lord's influence among these kings and their people, we must first consider how the Ptolemies came to power in Egypt. The story of Ptolemaic rule in Egypt begins with Aristotle's famous student, Alexander the Great, and his military conquests.

Alexander (356–323 BC) rose to power as the ruler of the Greek-Macedonian Empire following the death of his father, Phillip II, in 336 BC. After eliminating opposition in his homeland of Greece, Alexander embarked on an ambitious military campaign to conquer eastern lands ruled by King Darius III's Persian Empire. In just a few years, Alexander's forces conquered western Asia (modern day Turkey), Syria, and Egypt.

The Greek-Macedonian hegemony which spread over the known world with lightning speed was short-lived. After the death of Alexander in 323 BC, the empire split into three smaller kingdoms. Something that did last, however, was the Greek cultural and intellectual influence that Alexander's conquests brought to the known world. The mixing of Greek and Eastern culture that followed Alexander's conquests came to be known as the Hellenistic era. (Hellenistic means "like but not quite Greek.") The Hellenistic era was a prosperous time for Greek scholarly achievement, especially in the Egyptian city of Alexandria, which was founded by Alexander in 322 BC.

Following Alexander's death, Alexandria and all of Egypt came under the rule of Alexander's Macedonian general, Ptolemy I Soter. Interestingly, Ptolemy I (367–283 BC), his son Ptolemy II Philadelphus (308–246 BC), and his grandson Ptolemy III Eurgetes (280–221 BC) were scholars. Ptolemy I was a historian, Ptolemy II dabbled in zoology, and Ptolemy III was a mathematician. During their reign, these three Ptolemies constructed a library and museum at Alexandria to preserve literature and foster scientific activity. Alexandria's museum and library quickly became the world-capitol of learning where intellectuals gathered to work and study. Sprague De Camp described the significance of the Ptolemies' contributions as follows:

> In building the Library, the Ptolemies made a far greater contribution to civilization than all their palaces and parades. Many rulers have sought eternal fame: some by conquest and massacre, some by building grandiose temples and tombs, some by forcibly converting multitudes to their particular creed, and some by imposing a host of strangling rules and restrictions on their subjects. But few rulers have ever succeeded in doing so much good with so little suffering as the Ptolemies did in building up the Library of Alexandria.[11]

The Ptolemies' support of scholarly work resulted in considerable scientific progress at Alexandria. Notable examples include Euclid's (circa 300 BC) discoveries in geometry; Aristarchus' (circa 310 BC—230 BC) heliocentric (sun centered) model of the universe and attempts to measure astronomical distances; Archimedes' (circa 298 BC—212 BC) work in mathematics, geometry, and theoretical and fluid mechanics; and Eratosthenes' (circa

276 BC—195 BC) work in mathematics and geography which included estimating the earth's circumference with surprising accuracy.

A few centuries earlier, the prophet Daniel wrote that the Lord "setteth up kings: [and] giveth wisdom unto the wise, and knowledge to them that know understanding" (Daniel 2:21). I believe that these three scholarly-minded Ptolemies are a fulfillment of this scripture. Their efforts to improve mankind by building the library and museum at Alexandria suggest that they had wisdom, and their personal devotion to learning suggests that they were rulers who "knew understanding." Their wisdom and understanding qualified them to receive additional blessings of knowledge from the Lord.

The following story of King Philadelphus's (Ptolemy II) collaboration with seventy elders from Israel provides an uplifting example of the Lord giving the Ptolemies blessings of knowledge. According to the Jewish historian Josephus, a close friend of King Philadelphus named Aristeus, petitioned the king to release the Jewish slaves held captive in Egypt. Aristeus told the king that because the God who supported his kingdom was also the author of the Jewish law, and the Hebrews were devout worshipers of God, he should restore them to their home land in Palestine.[12] Upon hearing that the number of Hebrews in captivity exceeded 100,000, the king responded, "And is this a small gift that thou askest, Aristeus?" When others standing nearby replied that releasing the Hebrews would be a "thank offering . . . to that God who had given him his kingdom," Philadelphus was "much pleased" and ordered the release of the captives. He even went so far as to pay for the release of every Hebrew man, woman, and child enslaved to his soldiers.[13]

Now the chief librarian at Alexandria informed the king that despite their efforts to gather as many books as possible at the library, they had not yet obtained a record of the Jewish law. In an effort to procure a copy, Ptolemy II sent valuable gifts to Jerusalem to be dedicated to God, informed the Jewish high priest of the release of the Hebrew slaves in Egypt, and requested a copy of the Jewish law for the library in Alexandria. The high priest consented and sent seventy elders to Egypt to translate the law from Hebrew into Greek.[14]

When the seventy Jewish interpreters arrived in Alexandria, Ptolemy II asked to see their books containing the law. When the records were laid before him, the king, a collector and enthusiast of fine books, "stood admiring the thinness of those pages, and the exactness of the junctures, which could not be perceived (so exactly were they connected one with another); and this he did for a considerable time." After perusing the books for some time, he thanked the elders for coming, and gave "still greater thanks to him that sent them; and, above all, to that God whose laws they appeared to be." In reply and with one voice, the Jewish elders wished him happiness, an honor which, Josephus says, brought the king to tears.[15]

The interpreters were escorted down a causeway that extended into the Mediterranean Sea, and onto an island where, in quiet solitude, they carried out the work of translating the law from Hebrew into Greek. Every need and comfort was provided for so that the interpreters could translate without interruption. In all likelihood, the island location gave the interpreters a spectacular view of the newly constructed Great Lighthouse at Alexandria (Pharos), believed to have been completed during the reign of Ptolemy II.

When the translation was completed seventy-two days later, Ptolemy II admired the finished books "and gave order, that great care should be taken of them, that they might remain uncorrupted." He was also "delighted with hearing the laws read to him; and was astonished at the deep meaning and wisdom of the legislator [Moses]." Ptolemy II thanked the interpreters with personal gifts and valuables to be dedicated to God in Jerusalem. Furthermore, because he had gained such useful insights on the governance of mankind through his philosophical discussions with the Jewish interpreters, he sent an epistle to the high priest at Jerusalem requesting that the interpreters be allowed to visit him on occasion "because he highly valued a conversation with men of such learning."[16]

The resultant Greek translation of the Old Testament become known as the Septuagint, derived from the Latin word for seventy, *Septuaginta*, in recognition of the seventy elders who translated the work. The Septuagint, which is the oldest known biblical translation in the world, became the Bible of the early Christian church and the source of many Old Testament quotes that are now contained in our New Testament.[17] Certainly, this unusual collaboration between Jewish elders and third century BC Greek scholars was inspired by the Lord. Greek scholarship at Alexandria benefited from the wisdom of the Jewish elders, and the Greek translation of the Mosaic law (Septuaginta) preserved teachings of the ancient prophets for both early and latter-day Christians.

Subsequent Ptolemaic rulers in Egypt were not as concerned about scholarly endeavors as the first three Ptolemies. Waning interest in the library led to the gradual decline of scholarly work at Alexandria. Apathy toward the library was particularly evident during the reign of the second to last Ptolemaic ruler, Cleopatra VII. Cleopatra was apparently more interested in strengthening Ptolemaic rule in Egypt through political intrigue and deception than in scholarly endeavors. There is even evidence to suggest that she gave a large number of books from the library to Julius Caesar, with the intent of placating the Roman emperor to prevent him from invading Egypt. Unfortunately she gave the books to a civilization that was more interested in military conquests than scholarly pursuits. In the end, her priceless gift did not prevent Octavian Caesar Augustus from conquering Egypt in 30 BC.

Notes
1. Flavius Josephus, *Josephus: The Complete Works.* Translated by William Whiston, A.M. (Nashville: Thomas Nelson, 1998), 8.2.
2. For example, see Michael Grant's *The Rise of the Greeks* (New York: Barnes & Noble Books, 2005), 300. The author points out that such religious views did not become widely known to the Greeks from Persian sources until the fifth century BC.
3. Charles Freeman, *The Greek Achievement: The Foundation of the Western World* (Middlesex, England: Penguin Books, 1999), 159.
4. George Sarton, *A History of Science: Ancient Science Through the Golden Age of Greece* (New York: John Wiley & Sons, 1952), 265.
5. Ibid., 265.
6. Ibid., 266.
7. Ibid.
8. Ibid., 264.
9. Ibid., 271
10. Karl Popper, *Conjectures and Refutations* (New York: Routledge Classics, 2002), 200.
11. L. Sprague De Camp, *The Ancient Engineers: Technology and Invention from the Earliest Times to the Renaissance* (Barnes & Noble Books, 1993), 130-31.
12. Flavius Josephus, *Josephus: The Complete Works.* Translated by William Whiston, A.M. (Nashville: Thomas Nelson, 1998), 12.2.3.
13. Ibid., 12.2.4.
14. Ibid., 12.2.1–12.2.5.
15. Ibid., 12.2.11.
16. Ibid., 12.2.12–12.2.15.
17. *Illustrated Dictionary of the Bible.* Edited by H. Lockyer, Sr. (Nashville: Thomas Nelson, 1986), 166–67.

FOUR
EARLY CHRISTIANITY AND THE APOSTASY

Retrogression has come upon mankind because they have rejected the counsels and commandments of the Almighty. Advancement has come largely because men have been willing to walk, in part at least, in the light of divine inspiration.

—Joseph Fielding Smith

EARLY CHRISTIANITY

Two thousand years after the birth of our Savior, Apostles of the restored gospel of Jesus Christ described His life and mission as follows:

As we commemorate the birth of Jesus Christ two millennia ago, we offer our testimony of the reality of His matchless life and the infinite virtue of His great atoning sacrifice. None other has had so profound an influence upon all who have lived and will yet live upon the earth.

He was the Great Jehovah of the Old Testament, the Messiah of the New. Under the direction of His Father, He was the creator of the earth. "All things were made by him; and without him was not any thing made that was made" (John 1:3). Though sinless, He was baptized to fulfill all righteousness. He "went about doing good" (Acts 10:38), yet was despised for it. His gospel was a message of peace and goodwill. He entreated all to follow His example. He walked the roads of Palestine, healing the sick, causing the blind to see, and raising the dead. He taught the truths of eternity, the reality of our premortal existence, the purpose of our life on earth, and the potential for the sons and daughters of God in the life to come.

He instituted the sacrament as a reminder of His great atoning sacrifice. He was arrested and condemned on spurious charges, convicted to satisfy a mob, and sentenced to die on Calvary's cross. He gave His life to atone for the sins of all mankind. His was a great vicarious gift in behalf of all who would ever live upon the earth.[1]

So influential was the Savior's life and mission that the gospel message spread throughout the Mediterranean region. Apostles and missionaries labored diligently to "teach all nations, baptizing them in the name of the Father, and of the Son, and of the Holy Ghost" (Matthew 28:19). The Apostle Paul, in particular, experienced much success in bringing the gospel to the gentiles. He taught the gospel throughout Greek communities surrounding the Aegean Sea, specifically Asia (Ionia), Macedonia, and Achaia (Greek Peninsula). He also founded Christian churches in Greek cities like Ephesus, Philippi, Thessalonica, and Corinth.

As our Bible Dictionary points out, Paul's missionary success in these areas was partly due to Greek "civilization, culture, and philosophy."[2] Greek democratic institutions, rational analysis, and interest in truth claims that arose during the first secular illumination "were of great service to the Church."[3] When Paul taught the Greek gentiles the gospel, the people were not only free to accept the gospel of Jesus Christ, many were interested in the "new theology" because of its bold claims to truth. Also, because of their rational heritage many were prepared to follow Paul's reasoning. The Book of Acts states that while Paul and Silas were in Thessalonica, they entered into a synagogue where "they *reasoned with [the people]* out of the scriptures" (emphasis added) and taught them the mission and doctrine of Jesus Christ (Acts 17:2). A "great multitude" of devout Greeks who had gathered accepted Paul and Silas's reasoning, were touched by the Spirit, and believed (Acts 17:4).

Unfortunately, during this time many Greeks were still worshipping mythical gods. When Paul arrived in Athens he found the city full of idolatry (Acts 17:16), and after he healed a cripple in Lystra, the people engaged in idolatrous celebrations and made preparations to offer sacrifices to Jupiter (Barnabas) and Mercurius (Paul) (Acts 14:11–13). These incidents troubled Paul because many were not in tune with the rational tradition that had been established centuries earlier by scholars such as Socrates and Aristotle. Paul told the idolatrous people that God, "who in times past suffered all nations to walk in their own ways" (Acts 14:16), would no longer suffer them to seek after idols. They possessed a rational heritage that provided them with the means for finding God. If they would embrace that heritage and ponder the wonders of nature in a reasonable manner, they would discover for themselves that there is one true God of heaven and earth (Acts 14:17).

Many educated Greeks, including Epicurean and Stoic philosophers, were willing to listen to Paul because of their desire to "hear some new thing" (Acts 17:21)—a characteristic typical of scholarly-minded people. As Paul stood on Mars' hill, he taught the people about the Lord who made the heaven and earth and gives life to all living things. He told them that the Lord did not dwell in their temples nor in the idols their people had created. When he taught them about the Resurrection, some of the people became anxious and began mocking him, perhaps because the notion of being raised from the dead conflicted with their mystical belief in reincarnation. Yet some were intrigued by Paul's teachings and expressed an interest in hearing more, while others quickly accepted Paul's teachings and believed on his words.

There is an intriguing parallel between what Paul and Socrates taught the people of Athens. Centuries before Paul arrived in Greece, Socrates walked the streets of Athens, diverting people from their belief in idolatry by directing them toward a rational quest for truth. Looking back, it is clear that the intellectual tradition of Socrates and his followers prepared the Greeks and other gentiles to receive the gospel. When the Apostle Paul arrived four hundred years later, people were prepared to receive Paul's reasoning, and eager to discover the truth of the gospel. Moreover, because the Greek intellectual tradition spread beyond Greece's borders, other nations were similarly prepared to receive the gospel message. Thus we see the important role that the first secular illumination played in the conversion of the gentiles. Without a doubt it was inspired by the Lord to prepare the inhabitants of the earth for the gospel of Jesus Christ.

Near the end of his third mission to the Greeks, Paul called a group of elders from Ephesus to join him in Miletus. There he recounted his diligent efforts to teach the people, bore testimony, gave them counsel, and informed them that he would not be returning. Before boarding the ship for Judea, "he kneeled down, and prayed with them all" (Acts 20:36). Those who were with him "all wept sore, and fell on Paul's neck, and kissed him, Sorrowing most of all for the words which he spake, and that they should see his face no more" (v. 38).

What a moving sight this must have been, to see the first Christians among the Greeks kneeling near the shores of the Aegean Sea, demonstrating their faith in God, and showing their love and appreciation to the missionary who brought them the everlasting gospel. What makes this final gathering all the more interesting is that it took place in the same city where the first illumination began six centuries earlier. As mentioned previously, in 600 BC, Thales of Miletus set out on a scholarly journey that changed the intellectual landscape of the western world and prepared the gentiles for the gospel of Jesus Christ. The Lord works in mysterious ways.

APOSTASY

Following the death of Paul and the other disciples, apostolic authority was taken from the earth, and the Church of Jesus Christ fell into obscurity and darkness. During this time, mankind's receptiveness to the enlightening power of the Spirit of the Lord diminished. This diminishment affected mankind not only spiritually, but secularly as well. As the spiritual enlightening power of the Spirit of the Lord waned among mankind, so did the secular enlightening power. This concurrent loss of spiritual and secular enlightenment explains why there was a simultaneous occurrence of a spiritual apostasy and secular Dark Ages.

During the Apostasy mankind tried to compensate for the loss of apostolic authority and spiritual enlightenment by appealing to human reason. To this end, the theoretical contributions of ancient Greek scholars were monopolized by misguided theologians and scholars who hoped to settle doctrinal disputes and make Medieval Christianity more rationally appealing to the masses. The works of famous scholars like Plato and Aristotle now served the interests of a powerful apostate church. Furthermore, because the apostate church controlled a large number of political and educational institutions, few dared interpret Classical Greek principles in a manner contrary to the Church. As we shall see, this hijacking of Classical Greek philosophy by religious authority played a major role in driving the world deeper into spiritual and secular darkness.

Spiritual Decline

An early sign that the Apostasy was afoot was the rise of gnosticism in the late first century. Gnostics, as they were called, were largely concerned with esoteric and mystical explanations about an unknowable god, who we are, where we came from, and how we got here. Gnostic metaphysical speculations on these and other matters drew criticism from Christian traditionalists such as Irenaeus and Tertullian, who accused the gnostics of polluting church doctrine with philosophy and mysticism.

Despite the efforts of Irenaeus, Tertullian, and others to protect Christianity from outside influences, philosophy and mysticism eventually infiltrated Christian doctrine. Efforts to combine Christianity and philosophy surprisingly came from within.

The introduction of philosophy into Christianity was largely the work of scholars who, after converting to Christianity, combined religion and philosophy in an attempt to make Christianity more appealing to nonbelievers, especially the Romans who were persecuting Christians. One Greek scholar who devoted himself to this task was Clement of Alexandria (circa AD 150–215). As a theologian and head of the catechetical school of Alexandria in Egypt, Clement promulgated the belief that the Church would be held in higher regard and gain more converts if its beliefs were blended with ancient Greek philosophy. He therefore set out to "Hellenize" Christianity by combining Church doctrine with Greek philosophy, particularly that of Plato.

After Clement died, his student, Origen (circa 185–254 AD), continued the effort to infuse Church doctrine with Greek philosophy. Origen was a highly respected theologian and scholar in his day. Amid the confusion over doctrinal issues that prevailed during the third century, Origen observed that "there are many who profess to believe in Christ who disagree among themselves not only in small and minor matters, but rather about the great and the greatest matters."[4] These disagreements largely centered on the nature of the soul, resurrection, Holy Spirit, and incarnation of Christ. Origen believed that doctrinal truths could be discovered through intellectual study and reasoning, and thus set out to resolve doctrinal disagreements through personal study and theological discussions.

Origen's efforts to settle doctrinal disputes through study and discussion produced less than satisfactory results. Admitting defeat, he declared, "If anyone can find out anything better, or confirm by more evident proofs the assertions he makes concerning the Holy Scriptures, let such conclusions be accepted in preference to [mine]."[5] Thus we see how, as the spiritual apostasy was taking hold, "the foundations of doctrine had shifted from prophetic revelation to human reason."[6] This shift from prophetic authority to human reason was inevitable, given that priesthood authority to speak and act in the name of the Lord was taken from the earth. Without spiritual guidance from prophets and apostles, people began to rely on human reasoning to find answers to doctrinal questions.

Christian beliefs were gradually transformed as scholars and theologians infused church doctrine with philosophy. Nineteenth century Greek historian Edwin Hatch remarked, "It is therefore the more remarkable that within a century and a half after Christianity and philosophy first came into close contact, the ideas and methods of philosophy had flowed in such mass into Christianity, and filled so large a place in it, as to have made it no less a philosophy than a religion."[7] This influx of philosophy transformed many of the fundamental truths of the gospel into mystical and confusing doctrine.

Take, for example, the basic principle of faith. According to Hatch, "under the influence of contemporary Greek thought, the word Faith came to be transferred from simple trust in God to mean the acceptance of a series of . . . propositions in abstract metaphysics . . . concerning Him, His nature, relations, and actions."[8] Indeed this was a time when many plain and precious truths of the gospel were being lost (1 Nephi 13:26) and people were seeking the word of the Lord, but not finding it (Amos 8:13).

The transformation of fundamental Christian principles is especially evident in the outcome of the ecumenical council at Nicaea, in 325 AD. Emperor Constantine commissioned a council of theologians to settle a vigorous dispute over the nature of Jesus and his relationship to the Father. On one side of the debate were the followers of Arius, called Arians, who believed that Jesus was subordinate to and physically separate from the Father. On the other side were the followers of Athanasius who claimed that the Father and Son were physically one and the same being. After much theological debate on this issue, Emperor Constantine legitimized the position advocated by Athanasius. Arianism was declared heretical, and the medieval church officially adopted the doctrine known as homoousios ("of one substance"), which claims that the Father and Son are one and the same being.

Without the benefit of apostolic authority, the theological debate at Nicaea further distanced people from the truth about the nature of the Father and the Son. John A. Widtsoe stated, "This false doctrine [of homoousios], which has been nurtured though the centuries is an excellent illustration of philosophical-theological error and nonsense."[9] Hilary, who witnessed the outcome of the decision at Nicaea, observed the nonsense that the philosophical-theological quarreling had caused. He wrote, "Since each side is beginning to be [an] anathema to the other, it would seem that hardly anybody belongs to Christ (or is on Christ's side) any more. We are blown about by winds of doctrine, and as we teach we only become more upset, and the more we are taught, the more we go astray."[10] Efforts to maintain the doctrinal and intellectual authority of the medieval church by appealing to philosophy drove the people further into spiritual apostasy.

Christosophy—An Unholy Union

Attempts to create a rationally consistent religion by combining Christian doctrine and philosophy are endeavors I call christosophy. Medieval christosophy not only skewed humanity's understanding of the plain and simple truths of the gospel, it tied up the intellectual contributions of Plato and Aristotle in the service of apostate theology. Two influential scholars who attempted to systematize Christianity and Greek philosophy during the Dark Ages were Aurelius Augustine and Thomas Aquinas. Let's take a look at the detrimental effect their christosophies had on scholarly knowledge.

Aurelius Augustine (354–430 AD) received a Christian upbringing from his mother. He turned his back on his religious education when he went to Carthage to learn rhetoric, which, at the time, was required for a career in politics and law in the Roman Empire. Later on, while teaching at Milan, he experienced a life-altering conversion that brought him back to Christianity. He subsequently devoted his life to Christianity and joined the professional clergy in North Africa. But Augustine never completely abandoned his affinity for secular, pagan scholarship. While in Milan he also discovered Neoplatonism, a belief system that combined Christian mysticism and Plato's philosophy. He eventually integrated Neoplatonism and Christian doctrine in an attempt to put the latter on a rational foundation that appealed to the masses.[11]

Augustine's rational quest for God and truth adversely affected the intellectual atmosphere of medieval Europe for centuries. Georgetown University professor Daniel Robinson described the negative effect that Augustine's work had on science in the following way. Augustine's *Confessions*, Robinson wrote, "attacked intellectualism so broadly as to number too many casualties."[12] Also, Augustine's *City of God* "reduced that great interest in a daily civic life that is the mark of every truly 'classical' period,"[13] while his *De Libero Arbitrio* and *De Trinitate* "relentlessly drew attention and energy away from Stoic science, Aristotelian logic, and Platonic rationalism."[14] These circumstances led Robinson to conclude,

> To the extent that we consider an unornamented and disinterested search for truth to be a noble one and a positive enterprise, Augustine's influence must be judged harshly. His teachings and the eagerness and talents of his followers induced fear and humility of a sort antithetical to creativity and culture.[15]

Without a doubt, Augustine's christosophy created an anti-intellectual environment that contributed to the Dark Ages. His theology dominated the intellectual atmosphere of Europe until the thirteenth century when another influential theologian, Thomas Aquinas, introduced a new brand of christosophy that was similarly detrimental to creativity and scholarship.

Thomas Aquinas (1224–1274) was raised in the Christian faith and embraced religion at a young age. As a young man he enrolled in the Benedictine monastery at Monte Cassino. Against the urging of his family, he left Monte Cassino for the University of Naples, where he joined the Dominican order in 1244. Back then the Dominicans were required to get a university education and become skilled in the art of persuasion for the purpose of quelling heretical movements and winning the hearts and minds of the middle class, which was wavering in its allegiance to the papacy.

During the thirteenth century, a new threat to the authority of the papacy had arisen in Europe. That threat was the rediscovery of the complete works of Aristotle. People looking for an alternative to the neoplatonic Augustinian world-view began studying Aristotle's works. This new movement, which came to be known as Aristotelianism, created a crisis for the church. Aristotelian scholars such as the Averroists (named after the twelfth century Arabic philosopher Averroes, who wrote extensively on the works of Aristotle), were eager to point out inconsistencies between church doctrine and Aristotle's teachings.

Because of their education and expertise in quelling heretical movements, the Dominicans were mobilized in an effort to neutralize the threat posed by Aristotelianism. The "Dominicans perceived . . . that *in a reconciliation of Aristotle with orthodoxy lay the best hope of making Christianity intelligible to the new age.*"[16] The Dominican Thomas Aquinas dedicated himself to the task of synthesizing Aristotle's teachings with church doctrine. His brand of christosophy became known as Scholasticism.

There is some disagreement among historians over the impact that Aquinas's Scholasticism had on scholarship. On the positive side, Scholasticism helped disseminate Aristotelian ideas such as empirical observation and reason throughout the Latin west. Nevertheless, the dissemination of Scholasticism failed to improve scholarship because it tied up Aristotle's ideas in the service of institutionalized religion. According to historian B. R. Hergenhahn, "Once Aristotle's ideas were assimilated into church dogma, they were no longer challengeable. In fact, Aristotle's writings became almost as sacred as the Bible."[17] He added, "Although the scholastics were outstanding scholars and hairsplitting logicians, they offered little of value to either philosophy or psychology. They were more interested in maintaining the status quo than in revealing any new information."[18]

In conclusion, I think that Augustine and Aquinas had good intentions when they set out to merge philosophy with Christianity. Yet, at the same time, they must have known that their efforts were changing the original doctrines of the Christianity. As they added change upon change, the medieval church moved further away from the plain and precious truths of the gospel. And while they certainly had no intention of hindering scholarly work, their christosophies did just that. Hergenhahn summarized the effects of Augustine and Aquinas's work on scholarship when he wrote,

> During the time from Augustine to, and including, Aquinas, scholarship consisted of demonstrating the validity of church dogma. New information was accepted only if it could be shown to be compatible with church dogma; if this was not possible, then the information was rejected. The "truth" had been found, and there was no need to search elsewhere.[19]

DARK AGES

In the first few centuries following Rome's ascension to power (circa 30 BC), scholarship continued at Alexandria's library and museum on a limited scale. For instance, there were contributions from the first century mathematician and engineer Heron the "Hero of Alexandria," the second century mathematician and astronomer Claudius Ptolemy, and the third century mathematician Diophantus. But, by and large, original thinking and discovery at Alexandria and around the world gradually decreased during these first few hundred years as the spiritual apostasy came into full force.

This gradual loss of knowledge did not go unnoticed by scholars back then. Noting the changes taking place, Seneca, a Roman philosopher and personal assistant to Emperor Nero, declared with prophetic foresight that "the day will yet come when the progress of research through long ages will reveal to sight the mysteries of nature that are now wholly concealed. . . . The day will yet come when posterity will be amazed that we remained ignorant of things that will to them seem so plain."[20] He also declared that "far from advance being made toward the discovery of what the older generations left insufficiently investigated, many of their [past] discoveries are being lost."[21] Seneca's predictions came true. By the sixth century, the Bishop of Tours and historian Gregory of Tours lamented, "Woe to us, for the study of letters has disappeared from among us."[22] The world had fallen into darkness and would remain in darkness for more than a millennium.

The expression "Dark Ages" has recently come under scrutiny by some historians who argue that the Medieval Ages were not at all dark because of occasional, unique cultural achievements. Notwithstanding sporadic social and cultural progress, scholarly knowledge was in decline. The Dark Ages are generally believed to have lasted from AD 500–1000. However, if we refer to the beginning of the Dark Ages as the time when secular knowledge began to wane, and the end as the time when secular knowledge began to progress, then it can be argued that the Dark Ages began much earlier and lasted much longer.

In my opinion, the Dark Ages began a couple of centuries earlier than 500 AD, during the early third century AD, and lasted a few centuries longer than 1000 AD, until the fifteenth century. In other words, the Dark Ages began with the spiritual apostasy and ended with the Renaissance, when the Spirit of the Lord began enlightening mankind in preparation for the Restoration of the gospel of Jesus Christ. The lack of progress in two leading sciences of the Middle Ages, astronomy and human biology, supports this timeline.

Astronomy

The Greek mathematician Claudius Ptolemy (circa 100–165 AD) studied at Alexandria during the second century AD. His influential work, the *Almagest*, is a compilation of the astronomical and cosmological contributions of ancient Greek scholars. In the *Almagest*, Ptolemy espoused the Aristotelian geocentric model of the universe because objects in the heavens appear to rotate around the earth. He wrote, "If one should next take up the question of the earth's position, the observed appearances with respect to it could only be understood if we put it in the middle of the heavens as the centre sphere."[23]

The Ptolemaic model of the universe, as it came to be known, dominated cosmology throughout the Middle Ages. At the center lay the stationary earth composed of earth and water. Positioned immediately above the earth was a zone containing the primary element *Aer* (air), and beyond that was the zone containing the primary element *Fier* (fire). Meteor showers and comets were thought to be caused by random combustion in the zone of fire above the earth.

Beyond the fire zone lay the solid crystalline celestial spheres containing the planets and stars. The first celestial sphere was the moon. The rotation of the lunar sphere was thought to stir the fire beneath, which in turn stirred up the air, which churned the water on earth, resulting in an ever changing world. The next six outer crystalline spheres contained Mercury, Venus, the Sun, Mars, Jupiter, and Saturn, respectively. The eighth sphere, called the Firmament, contained the stars, and surrounding that was the *Primum Mobile* which is where God was thought to control and propel the lower spheres. Beyond the *Primum Mobile* lay the dwelling place of God.

An early problem with this model was the occurrence of retrograde planetary motion (i.e., the appearance of a planet temporarily moving backward as the earth's orbit overtakes the planet). In an attempt to account for this and other anomalies, scholars "corrected" the Ptolemaic model by adding new mechanisms and principles of operation. Some proposed that the planets inhabited more than one sphere, while Ptolemy added a complex series of epicycles which were secondary orbits the planets followed as they moved around the earth.

Despite mounting anomalies and increasing complexity, Ptolemy's geocentric-based model was relatively effective at predicting the positions of the stars, planets, and moon. In fact, geocentrism endured as the dominant cosmological model throughout the entire Middle Ages (approximately 1500 years). Evidence of its longevity may be found in the writings of the twelfth century bishop of Paris Peter the Lombard and the fourteenth century scholar Dante. Centuries after Ptolemy, these men described the heavens as being ordered into nine circular spheres, arranged in descending order from heaven down to the fixed earth in the center. Lombard and Dante also maintained that the point furthest from the dwelling place of God, namely the center of the earth, was the location of hell.[24]

Human Anatomy

Another example of stagnant scholarly progress during the Middle Ages comes from the human biological sciences. During the late second century, the Greek scholar Claudius Galen (131–200 AD) was considered by many to be the leading expert on human anatomy. He served as personal physician to four different emperors in Rome, where he had to rely on his experience working with gladiators and animals for much of his information on human anatomy because of Rome's ban on human dissection. Despite these limitations, he wrote several influential treatises on physiology, including his monumental work *On the Use of the Parts of the Human Body.*

Notwithstanding its inaccuracies, Galen's book remained the definitive work on human anatomy for nearly 1500 years. Up until the fifteenth century, many scholars unequivocally accepted Galen's work despite mounting evidence against some of its claims. Even during the sixteenth century, medical professors at the University in Padua read from Galen's work while performing dissections. Galen himself would not have approved of this unwavering acceptance of his work. He is believed to have emphasized the importance of original observation and cautioned people against relying too much on the past work of others.

Eastern Scholarship

The lack of scholarly progress during the Dark Ages was not limited to the western regions of the Roman Empire. Scholarly decline was also evident in the eastern half of the Roman Empire, which came to be known as the Byzantine Empire. While it is generally accepted that the Byzantines maintained a culture of intellectualism that surpassed the west, their scholarship rarely produced significant advancements. Like the west, Byzantine scholarship languished because of civil unrest, continual warfare, and "an unreasoningly deep reverence for the past which discouraged originality."[25] Moreover, Byzantine scholars such as John of Damascus tied up classical Greek philosophy in the service of theology, and Byzantine literature was largely devoted to hagiography which was fanciful writings on the lives of saints. These factors contributed to the Byzantine Empire's lackluster intellectual progress.

For a time, secular scholarship fared better farther to the east in Arabia. Islam came into contact with the works of ancient Greek and Hellenistic scholars during their conquest of the seventh and eighth centuries. Initially, the caliphs (Islamic rulers) were very supportive of Western scholarship. Manuscripts from the West were gathered and translated, centers of learning such as the House of Wisdom in Baghdad were erected, and intellectuals and scientists were patronized by Arabic rulers. These conditions led to advancements in mathematics (algebra and trigonometry), medicine (pharmaceuticals and compiling information), alchemy (distilling and synthesizing chemicals), and astronomy (producing star charts and reckoning times and seasons).

Notwithstanding these advancements, original contributions in Arabic scholarship were limited. This limited originality was particularly evident in medicine, where, "to the medical theories of Hippocrates and Galen, Islamic scholars added little,"[26] and in astronomy, where "they added few principles to those upon which Ptolemy and other Greeks had based their astronomical system."[27] Perhaps their greatest contribution was the translation and preservation of classical Greek philosophy and natural philosophy. Most notably, through the writings of Islamic philosophers such as Avicenna and Averroes, the West became reacquainted with the complete works of Aristotle during the twelfth century.

Eventually the Islamic East followed the Christian West into the Dark Ages. During the Middle Ages, Islam experienced a religious awakening that reacted harshly against science and philosophy. This awakening, combined with military incursions from Mongols in the East, drastically reduced Arabic efforts to preserve and promote secular scholarship.

Failed Attempts to Restore Scholarship

To their credit, some western rulers tried to restore scholarship in Europe. One noteworthy example is King Charlemagne. King Charlemagne ruled over most of Western Europe during the late eighth and early ninth centuries AD. He was a scholarly-minded king with a penchant for learning. After receiving letters from monasteries that were full of grammatical errors and "uncouth expressions,"[28] he became concerned about the education of the churchmen. He suspected that much of the clergy lacked the skill and wisdom necessary for understanding the scriptures.

To rectify this situation, he ordered bishops to expand their schools and libraries and to admit lay persons. He urged his people to "learn psalms, notes, singing, computus (arithmetic), [and] grammar," demanded that "the religious books that are given [to] them be free of faults," and asked that "care be taken that the boys do not damage them [books] either when reading or writing."[29] Charlemagne's noble efforts to restore scholarship and learning might have proven successful were it not for the civil war that broke out after his death as his sons struggled for power. Scholarship soon declined to the conditions that existed before he came to power.

Scholars were also unable to bring an end to the intellectual darkness. Two men who deserve mention for their valiant efforts to reform scholarship during the Middle Ages are Robert Grosseteste and Roger Bacon. Robert Grosseteste (circa 1175–1253) studied astronomy, mathematics, and physics at Oxford, where he used his influence as chancellor to encourage people to build upon the foundation laid down by Classical

Greek scholarship. Unfortunately, his efforts had little immediate effect.

Grosseteste's student, Roger Bacon (circa 1220–1292), a Franciscan monk, encouraged the universities to replace scholasticism with the study of mathematics, language, and the experimental sciences. He wrote a treatise (*A Compendium on the Study of Philosophy,* 1272) criticizing the theological and philosophical methods of the medieval scholastics as unproductive. When his patron pope Clement IV died, Bacon was severely chastised by the Franciscan order for his attacks on scholasticism. He was imprisoned for fifteen years and many of his treatises were destroyed. Fortunately, copies of Bacon's writings were preserved by a small group of monks who believed in his work. Although his work would help transform science centuries later, his efforts to reform education and encourage scientific activity produced little immediate effect. "Like so many other historical figures, he would be heard but not until later."[30]

Despite the efforts of scholars and rulers to promote intellectual work, the world remained in intellectual darkness. Why was this so? In *Greek Science after Aristotle,* G. E. R. Lloyd offers answers. He proposed that scholarly work may have waned because of (a) widespread skepticism regarding one's ability to uncover the causes of natural events, (b) the belief that ancient scholars possessed superior knowledge, (c) the lack of communication and manpower, and (c) the decline of agriculture.[31] "Yet," as Lloyd points out, "when we attempt to estimate the impact of these and other circumstances on the work of [scholars], we must admit that we are largely left to guesswork."[32]

Scholars are left guessing the cause of scholarly decline during the Dark Ages because they do not understand this basic fact, as expressed by Joseph Fielding Smith: "Retrogression has come upon mankind because they have rejected the counsels and commandments of the Almighty."[33] When the world rejected the Lord's gospel and fell into apostasy, it lost blessings associated with the enlightening power of the Spirit of the Lord. The consequence was intellectual darkness. Moreover, when apostolic authority was taken from the earth, mankind lost a power inextricably connected with truth. Speaking on the role of priesthood authority, John Taylor said, "It governs all things—it directs all things—it sustains all things—and has to do with all things that God *and truth* are associated with."[34]

By itself, mankind could not bring the world out of darkness. Ending the Dark Ages required an act of God. It required a second illumination from the outpouring of the Spirit of the Lord. This illumination came during the fifteenth century Renaissance and sixteenth century Scientific Revolution. Just as had happened in ancient Greece two millennia earlier, the Spirit of the Lord brought the inhabitants of the earth out of intellectual darkness and prepared them for the Restoration of the gospel of Jesus Christ.

Notes

1. "The Living Christ: The Testimony of the Apostles, The Church of Jesus Christ of Latter-day Saints," *Ensign*, Apr. 2000, 2.
2. See "Greece" in the Bible Dictionary.
3. Ibid.
4. Origen, Peri Archon I, 2, in PG 11:115–116.
5. Ibid., II, 8, 4–5, in PG 11:224.
6. Hugh Nibley, *The World and the Prophets*. 3d ed. (Salt Lake City and Provo: Deseret Book, Foundation for Ancient Research and Mormon Studies, 1987), 48.
7. Edwin Hatch, *The Influence of Greek Ideas on Christianity* (New York: Harper & Row, 1957), 125.
8. Ibid., 310–12.
9. John A. Widtsoe, *Evidences and Reconciliations* (Salt Lake City: Improvement Era, 1960), 58.
10. Hilary II, 5, in PL 10:566–567.
11. Hugh Nibley, *The World and the Prophets*. 3d ed. (Salt Lake City and Provo: Deseret Book, Foundation for Ancient Research and Mormon Studies, 1987), 89.
12. Daniel N. Robinson, *An Intellectual History of Psychology* (Madison, WI: University of Wisconsin Press, 1995), 80–81.
13. Ibid., 81.
14. Ibid.
15. Ibid.
16. Lynn D. White, *The Vitality of the Christian Tradition*. Edited by George F. Thomas (New York: Harper & Brothers), 106; emphasis added.
17. B. R. Hergenhahn, *An Introduction to the History of Psychology*. 3d ed. (Pacific Grove, CA: Brooks & Cole, 1997), 73.
18. Ibid., 75.
19. Ibid., 74–75.
20. L. Sprague De Camp, *The Ancient Engineers: Technology and Invention from the Earliest Times to the Renaissance* (Barnes & Noble Books, 1993), 254.
21. Ibid.
22. Joseph Dahmus, *A History of the Middle Ages* (Barnes & Noble Books, 1995), 120.
23. Michael Shermer, *The Borderlands of Science: Where Sense Meets Nonsense* (New York: Oxford University Press, 2001), 143.
24. Ibid., 144–47.
25. Joseph Dahmus, *A History of the Middle Ages* (New York: Barnes & Noble, 1995), 144.
26. Ibid., 188.

27. Ibid., 189.
28. Ibid., 206.
29. Ibid., 207.
30. Wayne Viney, *A History of Psychology: Ideas and Context* (Needham Heights: MA, 1993), 86.
31. G. E. R. Lloyd, *Greek Science After Aristotle* (New York: W. W. Norton & Company, 1973), 167–71.
32. Ibid., 167.
33. Joseph Fielding Smith, *The Way to Perfection* (Salt Lake City: Genealogical Society of Utah, 1949), 48.
34. John Taylor, *The Gospel Kingdom: Selections from the Writings and Discourses of John Taylor.* Selected, arranged, and edited, with an introduction by G. Homer Durham (Salt Lake City: Improvement Era, 1941), 128; emphasis added.

FIVE

THE SECOND GREAT ILLUMINATION

Have you ever tried to associate the outpouring of knowledge, the great discoveries and inventions during the past one hundred years, with the restoration of the Gospel? Do you not think there is some connection? It is not because we are more intelligent than our fathers that we have received this knowledge, but because God has willed it so in our generation!

—Joseph Fielding Smith

SCIENTIFIC REVOLUTION AND THE RESTORATION

The Apostle Paul prophesied that an outpouring of the Spirit of the Lord would take place in the last days. He said, "And it shall come to pass in the last days, saith God, I will pour out of my Spirit upon all flesh" (Acts 2:17). This outpouring of the Spirit ended the Dark Ages and prepared the inhabitants of the earth for the Restoration of the gospel of Jesus Christ. Commenting on the outpouring of the Lord's Spirit, Joseph Fielding Smith stated,

> This is the light spoken of by Joel, which has been poured out in these latter days and through which men are inspired to invent and discover the great truths which, until now, the Lord has seen fit to keep hid from the inhabitants of the world. We see the marvelous things which the Lord has revealed to man in this dispensation of the Fulness of Times, all in preparation for the restoration of all things.[1]

Bruce R. McConkie similarly declared,

> It was this spirit, the Light of Christ, which prepared the way for the opening of the dispensation of the fulness of times. Working in the hearts of men in the dark ages, it caused them to seek light, to translate the Bible, to break away (partially, at least) from the chains of religious darkness which bound their minds. This is the spirit, inspiring good men and honest truth seekers in every nation, which has led to the great discoveries, inventions, and technological advances of our modern civilization—achievements withheld from former dispensations and made known only in the last days.[2]

Signs of the Spirit illuminating people's minds were evident during the Renaissance. During the fifteenth century AD, the Spirit inspired two monumental discoveries that changed the world. There was Christopher Columbus's discovery of the Americas in 1492. When Nephi prophesied of this discovery more than two thousand years earlier, he wrote that the Spirit of the Lord "wrought upon the man [Columbus] and he went forth upon the many waters" (1 Nephi 13:12). Columbus acknowledged this guidance from the Spirit when he wrote, "With a hand that could be felt, the Lord opened my mind to the fact that it would be possible to sail from here to the Indies.... This was the fire that burned within me.... Who can doubt that this fire was not merely mine, but also of the Holy Spirit?"[3]

A second monumental discovery came during the 1450s when Johann Gutenberg developed movable type printing and printed copies of the Latin Bible. The Gutenberg Bible helped break the chains of ignorance by eroding clerical control over scriptural interpretation, making it possible for common people to read and interpret the Bible for themselves. Soon thereafter, religious reformers like Martin Luther began challenging apostate biblical interpretations and principles of the medieval church. The feeling of freedom that the Gutenberg Bible generated also liberated secular scholars from the intellectual restrictions imposed on them by dogmatic religious authority. Consequently, scholars increasingly challenged the church's authority on secular issues.

The scholarly freedom that ensued led to the publication of two monumental works in 1543. These two works further pierced the veil of darkness which hung over people's minds and launched an era of discovery that prepared the inhabitants of the earth for the Restoration of the gospel of Jesus Christ. One was a book on human anatomy, the other a book on astronomy.

In 1543, a young Flemish physician named Vesalius published a book titled *De Humani Corporis Fabrica (On the Structure of the Human Body)*.

This text corrected many of Galen's second century errors and contained the earliest accurate description of human anatomy. In the same year, a Polish scholar named Nicolaus Copernicus published *De Revolutionibus Orbium Coelestium (On the Revolutions of the Celestial Spheres)*. This text challenged Ptolemy's second century model of the heavens and presented a sun-centered model of the universe. These two works ushered in an era of rapid discovery called the Scientific Revolution.

So swift was the Spirit's illumination during the seventeenth century Scientific Revolution that "even with the finest tutors, a person could not keep up with all that was new."[4] Three inspired scholars who played a major role in this progress were Galileo Galilei, René Descartes, and Isaac Newton. Their unrelenting desire to reveal truth drove them forward despite opposition from powerful religious authority. Their philosophical and scientific contributions revolutionized the way we view our world, and by taking a stand against religious dogma, they weakened the church's control over spiritual and secular thinking.

These pioneers of the Scientific Revolution also helped prepare the earth for the Restoration of the gospel. In fact, their contributions were equally important to the contributions of the founding fathers of the American Constitution in terms of preparing the earth for the Restoration. The Lord said that the Constitution was established so that "every man may act . . . according to the moral agency which I [the Lord] have given unto him," and so that no "man should be in bondage one to another" (D&C 101: 78–79). As we shall soon see, the Lord brought about the Scientific Revolution to liberate mankind's views on religion, its relationship to God, and its ability to acquire knowledge. Moreover, just as the founding fathers of the Constitution encountered opposition in their attempts to liberate mankind from political tyranny, the pioneers of the Scientific Revolution encountered opposition in their attempts to liberate humanity from intellectual and religious tyranny.

We will begin by looking at Galileo Galilei's struggle to reveal truth about the solar system. Next, we'll review René Descartes's journey to uncover absolute truth about the existence of God, and how his rational approach empowered people to personally quest for truth. Finally, we'll take a look at Isaac Newton's remarkable journey of overcoming the odds to become one of the greatest scientific minds of all time. We'll also discuss how Newton's enlightened perspective on the Godhead nearly cost him his livelihood, and find out why he probably would have joined The Church of Jesus Christ of Latter-day Saints had he lived one century later.

GALILEO GALILEI (1564-1642)

Galileo was a key player in the divinely appointed secular illumination that took place during the seventeenth century. He had a strong desire to reveal truth. He exposed many falsehoods that had been accepted for centuries, such as the long-held Aristotelian belief that the speed of a falling body is proportional to its weight and that the heavens are unchanging. He also provided crucial evidence supporting the Copernican heliocentric model of the heavens.

Not everyone was excited about his discoveries, however. Many religious leaders who opposed his discoveries endeavored to discredit his work. Rather than back down from clerical pressure, he courageously pressed forward by confronting opposition and challenging entrenched false beliefs about the heavens and the church's exclusive right to interpret the Bible.

His brave challenge set the stage for a new era that looked to observation, experimentation, and personal discovery rather than ancient and clerical authority for answers. This challenge played an important role in freeing people's minds from the intellectual bonds imposed by the church. Because of these and other contributions, he is rightfully called the father of modern scientific thinking. But, as we shall see, this distinction came at a price; it cost him his freedom.

During the late sixteenth century, Copernicus's book *De Revolutionibus* did not pose a significant threat to the church-sponsored, earth-centered (geocentric) view of the universe. Most people who were aware of the book at that time simply regarded Copernicus's model as a convenient way to calculate the motion and position of planets. Also, when Copernicus's work was published, empirical evidence in favor of his heliocentric model was lacking. Thus his book was not immediately perceived as a threat to church authority and scholasticism; that was, at least until Galileo came along.

In 1609 Galileo learned of a Dutch lens maker who had developed an instrument which, when looked through, made distant objects appear closer. Galileo learned the principles by which the telescope was constructed, studied refraction, and then built his own. Said he, "First I prepared a tube of lead, at the ends of which I fitted two glass lenses, both plane on one side while on the other side one was spherically convex and the other concave. Then placing my eye near the concave lens I perceived objects satisfactorily large and near."[5] Galileo refined his design until, he says, "I succeeded in constructing for myself so excellent an instrument that objects seen by means of it appeared nearly one thousand times larger and over thirty times closer than when regarded with our natural vision."[6]

Apparently Galileo was the first to turn a telescope toward the sky and observe the sun, moon, stars, and planets. He was astonished with what he saw. Said he, "But forsaking terrestrial observations, I turned to celestial ones . . . with wondering delight."[7] He observed that the heavens were not perfect and unchanging as many ancient scholars and Scholastics had believed. He discovered, for instance, that the surface of the moon is rough and cratered like the earth, and that the sun has irregular spots that move. Notwithstanding his own skill and ingenuity, he gave thanks to God by declaring: "I render infinite thanks to God . . . for being so kind as to make me alone the first observer of marvels kept hidden in obscurity for all previous centuries."[8]

Galileo's telescope observations enabled him to begin building a case for the Copernican sun-centered (heliocentric) model of the solar system, a theory which he had always believed. A former student named Benedetto Castelli informed Galileo of a decisive test that would determine whether the planets revolved around a stationary sun. Castelli told Galileo that if Copernicus's theory was correct, then Venus, which most people believed was nearer to the sun than the earth, would go through phases like the moon as it orbited the sun. Galileo confirmed this hypothesis when he observed phases in Venus just as Castelli had predicted.

Galileo's observations of Venus's phases and Jupiter's moons composed compelling evidence supporting Copernicus's heliocentric model, yet many were not convinced. His telescope observations were disdained by those who viewed them as a threat to church-sponsored Aristotelian beliefs, the Bible, and the authority of the church. Some prominent theologians would not listen to him nor hear his evidence. Even a professor of philosophy at Padua University refused to look through Galileo's telescope for fear of what he might find. To make matters worse, during this time Europe was in the throes of the Counter-Reformation. Church leaders were lashing back at the Protestant Reformation and anyone who challenged their authority. They were censoring publications, placing discordant literary works on the *Index of Prohibited Books*, and using the Inquisition to investigate people who spread opposing viewpoints. With the support for heliocentrism steadily growing, the Inquisition set its sights on Galileo and Copernicus's book.

In 1616, Pope Paul V convened a meeting of theologians in Rome to reach a decision regarding Copernicus's heliocentric theory. Galileo went to Rome to meet with eminent church leaders and hopefully dissuade them from adopting a hostile position toward Copernicanism. But before Galileo could make his case, on March 5 the church council declared Copernicus's theory "false and completely contrary to the Divine Scriptures."[9] Copernicus's book *De Revolutionibus* was placed on the *Index of Prohibited Books* "until corrected."[10] The decree made no mention of Galileo or his findings; yet the church sent a clear message to him when, at the pope's request, Cardinal Bellarmine met with Galileo in private to warn him to not speak openly in favor of Copernicus.

Over the next few years, Galileo complied with the injunction to not openly advocate Copernicanism. However, his strong desire to teach the world the truth about the heavens persisted. In 1623 he saw an opportunity to publish his evidence supporting Copernicus's theory when an acquaintance named Maffeo Barberini was elected pope. Galileo met with Barberini (Pope Urban VIII) and secured permission to write about the Copernican theory on a hypothetical basis. Consequently Galileo wrote his *Dialogue Concerning the Two Chief World Systems*, a masterful fictional narrative comparing the merits of both geocentrism and heliocentrism. Unfortunately his book was perceived by some religious leaders as arguing too strongly in favor of Copernicanism, and to make matters worse, some clergy suggested that one of the fictional characters in the book was a disparaging caricature of the pope.

As a result, in the spring of 1633 Galileo was charged with violating the AD 1616 injunction prohibiting open support of Copernicanism. At seventy-four years of age, he was brought to trial, forced to make a confession renouncing the Copernican theory, and then placed under house arrest at his villa near Florence. Notwithstanding having lost his freedom, Galileo continued to write and conduct experiments while under house arrest. For example, during the last years of his life he wrote *Discourse on Two New Sciences*, a treatise on theoretical mechanics that laid the foundation for modern physics.

It is dismaying to see modern authors partially blame Galileo for what happened to him and insist that his attitude contributed to his demise. For instance, one author described Galileo as "a tactless and pugnacious self-publicist . . . who had a talent for making enemies."[11] This is an unfair characterization of Galileo. The accusation of "tactless" is not consistent with Galileo's lucid scholarly publications and willingness to share his evidence with others.

Also, he cannot be faulted for aggressively championing what he knew to be true and for criticizing self-proclaimed experts who spread falsehoods. His bold stance was most certainly a reaction against attacks by scholars who claimed the authority of the church and the Bible without considering scientific evidence. At every turn to make known the truth regarding the heavens, he was encumbered by efforts to silence him.

His relationship with children and friends reveals his kindness and concern for others. For instance, despite his busy lifestyle and having to cope with a recurring illness that confined him to his bed for long periods of time, he graciously assisted the convent at San Matteo where his two daughters resided. In letters to her father, Suor Maria Celeste repeatedly "thank[ed] him for some recent act of thoughtfulness or generosity toward herself, her sister, or someone else in the convent."[12] Dava Sobel, author of *Galileo's Daughter*, wrote,

> Thus, all the while that Galileo was inventing modern physics, teaching mathematics to princes, discovering new phenomena among the planets, publishing science books for the general public, and defending his bold theories against established enemies, he was also buying thread for Suor Luisa, choosing organ music for Mother Achillea, shipping gifts of food, and supplying his homegrown citrus fruits, wine, and rosemary leaves for the kitchen and apothecary at San Matteo.[13]

Galileo has also been accused of trying to embarrass the church by challenging its authority over the interpretation of biblical scripture, in particular, scripture which seemed to support the notion that the earth stands still and the sun moves.[14] Special attention was given to Joshua 10:12, which reads, "Then spake Joshua to the Lord in the day when the Lord delivered up the Amorites before the children of Israel, and he said in the sight of Israel, Sun, stand thou still upon Gibeon; and thou, Moon, in the valley of Ajalon." Galileo reasoned that because God is the author of both nature and the Bible, truths found in nature could not contradict truths found in the Bible. He suggested that scriptures like Joshua 10:12 were written to accommodate people's level of understanding and therefore should not be taken literally. He said,

> But since [Joshua's] words were to be heard by people who very likely knew nothing of any celestial motions beyond the great general movement from east to west, he stooped to their capacity and spoke according to their understanding, as he had no intention of teaching them the arrangement of the spheres, but merely of having them perceive the greatness of the miracle.[15]

Galileo's understanding of the scriptures may have violated the 1546 ruling by the Council of Trent (Session IV) which decreed that the interpretation of scripture was limited to bishops and church councils. Yet, as one who "held scripture and its message of redemption in high regard,"[16] Galileo clearly had no intention of embarrassing the clergy by challenging their interpretation of biblical scripture.

Moreover, Galileo's assertion that the scriptures were written to accommodate people's level of understanding agrees with the gospel principle that the Lord "speaketh unto men according to their language, unto their understanding" (2 Nephi 31:3), and that He reveals things to His servants "in their weakness, after the manner of their language, that they might come to understanding" (D&C 1:24). Galileo's concern over the correct interpretation of biblical scripture imitates our eighth article of faith, which declares that "We believe the Bible to be the word of God as far as it is translated correctly."

Galileo's experience of sharing his message of heliocentrism with religious leaders bears some resemblance to the Prophet Joseph Smith's experience of sharing the account of the First Vision with clergy. Perhaps this should come as no surprise given that they both challenged deep-seated religious views. The Prophet Joseph Smith declared, "I soon found that my telling the story had excited a great deal of *prejudice against me among the professors of religion*, and was the cause of great persecution" (JS—H 1:22; emphasis added). Galileo similarly wrote, "I discovered in the heavens many things that had not been seen before our own age. The novelty of these things, as well as some consequences which followed from them . . . *stirred up against me no small number of professors*."[17]

After sharing the First Vision, Joseph Smith wrote that certain clergy members attempted to *deny and disprove it* by claiming that "there were no such things as visions or revelations in these days; that all such things had ceased with the apostles and that there would never be any more of them" (JS—H 1:21). In like manner Galileo recounted, "Showing a greater fondness for their own opinions than for truth, [theologians] sought to *deny and disprove* the new things which, if they had cared to look for themselves, their own senses would have demonstrated to them."[18]

A level-headed person might have relented from the pressure, especially given the punitive actions of the Inquisition; this was a time when people were burned at the stake for openly espousing heretical views. But Galileo pressed forward undaunted, leading us to ask, "Why did he risk loss of life and liberty trying to convince the world of the truthfulness of the Copernican model of the heavens?" The answer is he was inspired by the Spirit of the Lord.

His work served a purpose higher than popularizing observation and mathematics in science. His work was instrumental in breaking the veil of ignorance that constrained people's minds. He laid the groundwork for an intellectual revolution wherein people were free to think for themselves. His work also encouraged people to personally seek out answers to scientific and spiritual questions, and to reevaluate mankind's place in the universe and relationship to God. These conditions were essential to the forthcoming Restoration of the gospel.

RENÉ DESCARTES (1596–1650)

René Descartes also played an important role in preparing the earth for the restored gospel. His epistemological contributions on how we can acquire truth revolutionized the intellectual atmosphere of the western world. Many of his ideas, which earned him the title "father of modern philosophy," are presented in influential treatises such as *Meditations on First Philosophy*, *Principles of Philosophy*, and *Passions of the Soul*.

Descartes challenged the medieval "scientific method" known as Scholasticism. He believed that scholasticism was an unprofitable enterprise, and endeavored to remove it from its place of prominence as the main mode of inquiry. *History of the Royal Society*, a seventeenth century account of the events surrounding the formation of the British Royal Society, describes the inept approach of the medieval scholastics. It records that the Scholastics "began with some generall Definitions of the things themselves, according to their universal Natures: Then divided them into their parts, and drew them out into severall propositions, which they layd down as Problems."[19] The problems they deduced then became the focus of debate as "they controverted on both sides: and by many nicities of Arguments, and citations of Authorities, confuted their adversaries, and strengthned their own dictates."[20]

In other words the scholastics' truth claims were grounded in appeals to authority, rhetoric, and eloquent arguments. This method "was never able to do any great good towards the enlargement of knowledge."[21] Rather, it "did more hurt, th[a]n good: and [only] serv'd to carry them the farther out of the right way."[22]

Descartes set out to discover a better way of acquiring knowledge. Reason played a central role in his approach; for him it was a major source of truth. He described his reason-based approach in his *Discourse on Method for Conducting One's Reason Well, and for Seeking the Truth in the Sciences.* This treatise is influential not only for the rational epistemology it introduces, but for its focus on the existence of God and the human soul (spirit).

Chapter four of his *Discourse on Method* describes how, in his search for a metaphysical certainty, Descartes began by doubting the reality of his thoughts, which he imagined were mere illusions like thoughts in a dream. He also doubted the reality of his sensations which, he claimed, could be deceiving him. He wrote, "Because I then desired to devote myself exclusively to the search for the truth, I thought it necessary that . . . I reject as absolutely false everything in which I could imagine the least doubt, in order to see whether, after this process, something in my beliefs remained that was entirely indubitable."[23]

After successfully doubting the existence of all that he experienced as real, he came to the conclusion that he could not doubt the fact that he doubted. Said he,

> I noticed that, while I wanted thus to think that everything was false, it necessarily had to be the case that I, who was thinking this, was something. And noticing that this truth—*I think, therefore I am*—was so firm and so assured that all the most extravagant suppositions of the skeptics were incapable of shaking it. . . . From the very fact that I thought of doubting the truth of other things, it followed very evidently and very certainly that I existed.[24]

Descartes described the "I" as the thinking part of a human being that is different from the physical body. It is, as he claimed, "the soul through which I am what I am, is entirely distinct from the body and is even easier to know than the body, and even if there were no body at all, it would not cease to be all that it is."[25] By this process he claimed to prove the existence of the soul.

Descartes also reasoned that, because he was imperfect, he could not be wholly responsible for thoughts of perfection. Thoughts of perfection had to come from a source that was perfect. "It is," he said, "no less a contradiction that something more perfect should follow from and depend upon something less perfect than that something should come from nothing."[26] He therefore concluded that these thoughts of perfection were "placed in me by a nature truly more perfect than I was and that it even had within itself all the perfections of which I could have any idea, that is to say, to explain myself in a single word, that it was [of] God."[27]

Descartes believed that God is the source of mankind's ability to rationally discern truth. He claimed that deity endows us with powers of reason that are capable of discerning truth and that "all our ideas or notions must have some foundation of truth, for it would not be possible that God, who is all-perfect and all-truthful, would have put them in us without that."[28] He also believed that the things we "very clearly and very distinctly" conceive of as being real are true because God lives, He is a perfect being, and "all that is in us comes from him."[29] Thus we can trust our clear and distinct perceptions of truth because a perfect creator who endowed us with the powers of reason would not allow us to be deceived when we clearly and distinctly perceive.

Descartes's philosophy on obtaining knowledge played an important role in shifting mankind's focus away from the unproductive method of the scholastics. His contributions gave people hope in a more reliable approach to acquiring knowledge about the world. He declared,

> It is possible to arrive at knowledge that would be very useful in life and that, in place of that speculative [scholastic] philosophy taught in the schools, it is possible to find a practical philosophy, by means of which, knowing the force and actions of fire, water, air, the stars, the heavens, and all the other bodies that surround us, . . . we might be able . . . to use them for all the purposes for which they are appropriate . . . to render men generally more wise and more adroit than they have been up until now.[30]

By pointing mankind in a better direction for discovering truth, Descartes played an important role in preparing the inhabitants of the earth for the restored gospel. He popularized the notion that it was no longer necessary to rely on the wisdom of the ancients or clergy (scholasticism) for answers to questions about God and the world. Ordinary individuals could discover knowledge themselves through faith, study, and reason, and feel assured that the Lord would not lead them astray. This contribution shaped the intellectual atmosphere of the west, which in turn influenced the truth-seeking pioneers of the Restoration such as Joseph Smith, Parley P. Pratt, and John Taylor.

Interestingly, Descartes was not going to publish his *Discourse on Method* after observing the church's hostile reaction to Galileo's *Dialogue*. He feared that the church might respond harshly to his work as well. He wrote that the fear of reprisal "was sufficient to make me change the resolution I had had to publish my opinions."[31] However, he decided to publish when he realized that he "could not keep [his ideas] hidden away without sinning grievously against the law that obliges us to procure, as much as is in our power, the common good of all men."[32] Clearly, he feared God more than man and sensed that his work was important in fulfilling the purposes of a higher power.

ISAAC NEWTON (1643-1727)

The culminating figure in the second illumination was Isaac Newton. Newton is often referred to as the greatest scientific mind of all time because of his contributions in calculus, optics, physics, astronomy, and for changing the way we understand the world. Some of his contributions are contained in his monumental book, *The Mathematical Principles of Natural Philosophy*, or *Principia* for short. Many scholars have called *Principia* the most influential science book ever written.

Few people realize that Newton was equally dedicated to theology. He was an avid student of the Bible, especially scripture dealing with the latter days. Because some of his religious views contradicted the tenets of the church which dominated political and educational institutions in seventeen to eighteenth century England, Newton had to keep his theological ideas secret.

Nowadays many scholars view Newton as a somewhat heretical and misguided genius when it comes to theological issues. But, as we shall see, when his life and accomplishments are viewed within the perspective of the restored gospel, we discover the true man—a man of integrity and unsurpassed intellect who was inspired by the Light of the Lord. His contributions played a major role in advancing knowledge and preparing the inhabitants of the earth for the Restoration of the gospel.

Given his intellectual accomplishments, one might think that Isaac Newton was born into privileged circumstances and received an upper-class education. In fact, the opposite is true. He was born premature and frail in 1643 (one year after Galileo's death) in a family farmhouse and was not given very good chances of survival. To make matters worse, his father and family provider, Isaac Newton Sr., died shortly before he was born.

As a young boy, Isaac was lonely and experienced challenges associated with being small for his age. When he was just three years old, his mother, Hannah Newton, accepted an offer to marry a man twice her age named Barnabas Smith. Because of an agreement between his mother and Barnabas, young Isaac was not permitted to live with his mother and stepfather. Instead he was left at the family farmhouse in his grandmother's care.

Isaac was reunited with his mother when his stepfather died in 1653, but the reunion was short-lived. Upon his return home, Isaac's mother wanted him to go to school, but the Puritan school in Grantham that he was to attend was too far away for him to walk, so his mother arranged for him to board with a chemist near the school. Separated from his mother once again, Isaac felt lonely and abandoned. At twelve years of age he described his feelings as such: "A little fellow; My poore help; Hee is paile; There is no room for me to sit; In the top of the house—In the bottom of hell; What imployment is he fit for? I will make an end. I cannot but weepe. I know not what to doe."[33] The boy who would become one of the greatest scientific minds of all time was in the depths of despair.

While boarding at the house of the chemist, William Clarke, and attending school in Grantham, Isaac began demonstrating unique ability in science and math. He developed a penchant for reading and handwriting copies of scholarly works in theology, astronomy, alchemy, and mathematics. At the apothecary he experimented with Mr. Clarke's chemicals and learned to boil, mix, and grind substances with mortal and pestle. He also spent his own time studying the mechanical principles of pulleys, levers, and gears and is known to have built a water clock and small windmills and watermills. He also liked to have fun. He built kites with lanterns and flew them at night to frighten the local residents.

When he turned sixteen, his mother brought him home to work on the family farm. This arrangement did not last long, for Isaac preferred reading and studying to watching livestock. One day he was so preoccupied with his thoughts about nature that the sheep he was tending wandered off and trampled a neighbor's barley, an infraction for which he was fined. Some people must have thought that Isaac was a lazy lad, but others recognized his burgeoning genius. His uncle, who was then rector of Trinity College at the University of Cambridge, and his Grantham school teacher intervened on Isaac's behalf. They convinced his mother to allow him to attend university. In June 1661, Isaac Newton was admitted to Cambridge at age eighteen.

University life did not bring an end to Newton's challenges. The curriculum in seventeenth century Cambridge was still dominated by medieval scholasticism. Furthermore, because his mother had given him little money, he had to enroll as a sizar, the lowest status attainable by a student. As a sub-sizar, Newton had to make a living performing menial services for other students such as "running errands, [and] waiting on them at meals."[34] He occasionally satisfied his hunger by "eating their leftovers."[35] Notwithstanding these and other challenges, he excelled at learning. Through self-study he pushed his knowledge beyond what was in the scholarly texts and possessed by university professors. He overcame enormous obstacles to become a leading scientist and scholar.

Newton was blessed with amazing God-given talents of perception and analysis. Newton biographer James Gleick described his perceptual abilities as follow.

> When [Newton] observed the world it was as if he had an extra sense organ for peering into the frame or skeleton or wheels hidden beneath the surface of things. He sensed the understructure. His sight was enhanced, that is, by the geometry and calculus he had internalized. He made associations between seemingly disparate physical phenomena and across vast differences in scale. When he saw a tennis ball veer across the court at Cambridge, he also glimpsed invisible eddies in the air and linked them to eddies he had watched as a child in the rock-filled stream at Woolsthorpe. When one day he observed an air-pump . . . creating a near vacuum in a jar of glass, he also saw what could not be seen, an invisible negative: that the reflection on the inside of the glass did not appear to change in any way. No one's eyes are that sharp. Lonely and dissocial as his world was, it was not altogether uninhabited; he communed night and day with forms, forces, and spirits, some real and some imagined.[36]

From his own writings, we also know that Newton was inspired by the Spirit of the Lord. He had an unquenchable passion to reveal secular and spiritual truth. "Truth," he declared, was "the offspring of silence and meditation."[37] Describing his personal approach to revealing truth, he said, "I keep the subject constantly before me and wait 'till the first dawnings open slowly, by little and little, into a full and clear light."[38] Such was the case when he discovered the universal law of gravity "by thinking on it continually."[39]

I find it interesting that he described his discoveries as "light." I think that he was keenly aware of the resemblance between truth and light, in the same way that D&C 84:45 declares that "whatsoever is truth is light." This scripture also tells us that "whatsoever is light is Spirit, even the Spirit of Jesus Christ," suggesting that Newton's ultimate source of light was the Lord. I believe that he knew that his inspiration came from the Lord, for he was a devout Christian and gifted theologian.

For Newton, a resplendent and orderly solar system evidenced a supreme creator. He stated, "This most beautiful system of the sun, planets, and comets, could only proceed from the counsel and dominion of an intelligent and powerful Being."[40] He also believed that the creator governs all things, is in and through all things, and is actively involved in

His creations. In fact, Newton hoped that his physical principles of the universe would help nonbelievers come to an understanding of the Lord. Regarding the *Principia*, he stated, "When I wrote my treatise about our Systeme I had an eye upon such Principles as might work w/th considering men for the beleife of a Deity & nothing can rejoyce me more then to find it usefull for that purpose."[41]

Newton had strong convictions about the nature of the Godhead. From his study of the Bible, he firmly believed that the Father and Son are separate individuals, and that the Son is submissive to the Father. He was also convinced that homoousious, the doctrine that the Father and Son are one and the same being, is false. Confusion over the nature of the Godhead, he believed, largely resulted from the intentional altering of biblical scripture by depraved priests and rulers who supported homoousious.

Newton's beliefs about the Godhead jeopardized his position as Lucasian Professor of Mathematics at Cambridge. Doctrines opposing homoousious had been declared heretical by most religious organizations, including the Trinity College of Cambridge where he worked. In 1675 he was to take an oath affirming Anglican doctrine and be ordained to the ministry. Rather than take an oath that contradicted his religious convictions, he decided to resign his prestigious position. At the last moment, King Charles II intervened and declared that the Lucasian professorship was exempt from having to take holy orders, thus enabling Newton to retain his post at Cambridge.

Because he rejected homoousious, many criticize Newton for rejecting the divinity of the Savior. In the eyes of those who accept Trinitarianism, separating the Father and the Son is akin to denying the divinity of the Son because, as they suppose, there is only one God. Nevertheless, restored gospel teachings illustrate that Newton was correct in believing that the Father and the Son are physically separate individuals.

Newton also believed, as the gospel teaches, that although there are many gods, for us there is but one Father and one Lord Jesus Christ, who is the Redeemer of all mankind and Creator of heaven and earth. The following statement by Newton captures this doctrine with surprising clarity, suggesting that his beliefs were anything but heretical. He wrote,

> We must beleive that there is one God or supreme Monarch that we may fear & obey him & keep his laws & give him honour and glory. We must beleive that he is the father of whom are all things, & that he loves his people as his children that they may mutually love him & obey him

as their father. We must beleive that he is the [universal ruler], Lord of all things with an irresistible & boundless power & dominion that we may not hope to escape if we rebell & set up other Gods or transgress the laws of his monarchy, & that we may expect great rewards if we do his will. . . . For tho there be [many] that are called god whether in heaven or in the earth (as there are Gods many & Lords many), yet to us there is but one God the father of whom are all things & we in him & one Lord Jesus Christ by whom are all things & we by him; that is, but one God & one Lord in o/r worship.[42]

Perhaps most interestingly, Newton believed that the basic doctrines of the early Christian church had been lost or distorted and that the true church of Jesus Christ was not upon the earth. He hoped to restore the truths and doctrines of early Christianity but gave up this task later on in life. His close associate and replacement as Lucasian professor at Cambridge, William Whiston, stated that Newton gave up all hope for a restoration of ancient Christianity because his "interpretation of the prophecies led [him] to expect a long age of corruption before [the Restoration] would take place."[43]

Newton was right with regard to the long night of spiritual darkness, but what he did not realize was that the Apostasy was nearly over. The Restoration of the gospel that he longed for was less than a century away. Ninety-three years following his death in 1727, the windows of heaven were finally opened and the fulness of the gospel of Jesus Christ was restored to the earth through the prophet Joseph Smith.

During the early years of the LDS Church, thousands of people who had been searching for the truth in England joined the Church after being taught by missionaries like Wilford Woodruff. If Isaac Newton had been alive to hear their gospel message, I believe that this great man would have recognized the doctrines of the gospel of Jesus Christ and possibly have been baptized.

CONCLUSION

Notwithstanding scholarly mistakes and personal foibles, Galileo, Descartes, and Newton were inspired participants in the second secular illumination. Their contributions were instrumental in bringing the world out of intellectual darkness and preparing the inhabitants of the earth for the Restoration of the gospel. According to Palmer and Colton's *A History of the Modern World,*

> The scientific revolution of the seventeenth century had repercussions far beyond the realm of pure science. It changed ideas of religion and of God and man. And it helped to spread certain . . . beliefs such as that the physical universe in which man finds himself is essentially ordered and harmonious, that the human reason is capable of understanding and dealing with it, and that man can conduct his own affairs by methods of peaceable exchange of ideas and rational agreement. Thus was laid a foundation for belief in free and democratic institutions.[44]

In other words, by challenging deep-seated, wayward religious views about God, man, and the universe, Galileo, Descartes, and Newton fostered an environment of intellectual freedom wherein people could discover truth for themselves rather than relying on religious and ancient authority. Intellectual freedom was necessary for the Restoration to succeed; people had to feel free to openly question religious doctrine, seek out religious truth for themselves, and share their knowledge with others. Without a doubt, these men played an important role in preparing the earth for the Restoration of the gospel.

As pointed out previously, a few centuries after the first illumination in ancient Greece, mankind fell into spiritual and secular darkness because of apostasy. Fortunately this has not been the case following the second illumination. Knowledge will continue to progress in the latter-days because the fulness of the everlasting gospel will never be taken from the earth again. The Spirit of the Lord will continue to be poured out upon the inhabitants of the earth just as Paul testified two thousand years ago.

Notes

1. Joseph Fielding Smith, *Man, His Origin and Destiny* (Salt Lake City: Deseret Book Co., 1954), 66.
2. Bruce R. McConkie, *Mormon Doctrine*. 2d ed. (Salt Lake City: Bookcraft, 1966), 716.
3. Arnold K. Garr, *Christopher Columbus: A Latter-day Saint Perspective* (Provo: BYU Religious Studies Center, 1992), 3.
4. L. Pyenson and S. Sheets-Pyenson, *Servants of Nature: A History of Scientific Institutions, Enterprises, and Sensibilities* (New York: HarperCollins, 2000), 48.
5. *Discoveries and Opinions of Galileo.* Translated by Stillman Drake (New York: Doubleday, 1957), 29.
6. Ibid.
7. Ibid.
8. Dava Sobel, *Galileo's Daughter* (New York: Penguin Books, 2000), 6.
9. Richard Blackwell, *Galileo Galilei, in Science and Religion: A Historical*

Introduction. Edited by Gary Ferngren (Baltimore, Maryland: John Hopkins University Press, 2002), 112.
10. Ibid.
11. John Henry, *Moving Heaven and Earth: Copernicus and the Solar System* (Cambridge: Totem Books, 2001), 94–95.
12. Dava Sobel, *Galileo's Daughter*, (New York: Penguin Books, 1999), 118.
13. Ibid., 118–19.
14. John Henry, *Moving Heaven and Earth: Copernicus and the Solar System* (Cambridge: Totem Books, 2001), 94–95.
15. *Discoveries and Opinions of Galileo*. Translated by Stillman Drake (New York: Doubleday, 1957), 211.
16. Ian Barbour, *Religion and Science: Historical and Contemporary Issues* (San Francisco, CA: Harper, 1997), 15.
17. *Discoveries and Opinions of Galileo*. Translated by Stillman Drake (New York: Doubleday, 1957), 175; emphasis added.
18. Ibid; emphasis added.
19. Thomas Sprat, *History of the Royal Society* (St. Louis, MO: Washington University, 1958), 16.
20. Ibid.
21. Ibid.
22. Ibid., 18–19.
23. René Descartes, *Discourse on Method*. 3d ed. Translated by Donald Cress (Indianapolis, Indiana: Hackett Publishing, 1998), 18.
24. Ibid; emphasis added.
25. Ibid., 19.
26. Ibid.
27. Ibid.
28. Ibid., 22.
29. Ibid., 21.
30. Ibid., 35.
31. Ibid., 34.
32. Ibid., 34–35.
33. James Gleick, Isaac Newton (New York: Pantheon Books, 2003), 14.
34. Ibid., 21.
35. Ibid.
36. Ibid., 90.
37. Ibid., 38.
38. Ibid.
39. Richard Westfall, *The Life of Isaac Newton* (Cambridge: Cambridge University Press, 1994), 40.
40. Ibid., 290.
41. Ibid., 205.
42. Ibid., 303–304.
43. Ibid., 302.
44. R. R. Palmer and J. Colton, *A History of the Modern World*. 3d ed. (New York: Alfred Knopf, 1965), 260.

SIX

DREAMS AND INSPIRATIONS

Blessed is the man who carries within him a God, an ideal, and who obeys it—ideal of art, ideal of science, ideal of gospel virtues. Therein lies the springs of great thoughts and great actions.

—Louis Pasteur

The Lord has been pouring out His Spirit upon the earth during the last days. This blessing is the main reason for the rapid progress in secular knowledge that began during the scientific revolution and continues to this day. A few of the ways in which the Spirit blesses mankind with secular knowledge are found in Acts 2:17, which reads: "Your sons and daughters shall prophesy, and your young men shall see visions, and your old men shall dream dreams." Evidence of these spiritual manifestations can be found in historical accounts of dreams, premonitions, and impressions given to those who qualified to receive such marvelous blessings.

As Latter-day Saints, we tend to associate the "marvelous work" that the Lord is performing among the children of men (1 Nephi 22:8) with the spiritual blessings of the restored gospel. But "marvelous work" also refers to the secular knowledge He has provided mankind. It is important for us to recognize the marvelous secular knowledge He has provided, for He has instructed us to "make known his wonderful works among the people" (D&C 65:4). For this reason, I shall share a few wonderful discoveries made possible by the Spirit of the Lord.

To begin with, we'll take a look at the inspiration that led to a monumental breakthrough in human biology, the discovery of the neuron.

Next, we'll consider the discovery of penicillin, a blessing without which twenty percent of you might not be reading this book—you, your parents, or your grandparents would have succumbed to an infectious disorder. Then we'll look at dreams in organic chemistry that unlocked secrets of biological life. Following this, we'll review the inspiration that led to the discovery of the cause of malaria and saved countless lives, and we'll examine how inspiration from the Spirit of the Lord led to the invention of the Periodic Table and the discovery of elements previously unknown to mankind. Finally, we'll look at how the light and hope of the restored gospel reduced the number of deaths from disease during the late nineteenth century.

THE NEURON

The single-cell neuron is the fundamental building block of the nervous system. The human nervous system is made up of millions of neurons, all organized in an intricate manner that enables us to sense our environments and carry out sophisticated motor and cognitive functions with amazing efficiency. The discovery of the neuron was made possible by nineteenth century scientists who sought to uncover the shape and function of mysterious microscopic filamentous-looking structures in the nervous system. This discovery did not come easily, however; it took the persistence of one scientist and the blessings of heaven to unlock the mysteries of the neuron.

In 1873 an accidental spillage led Camillo Golgi (1843–1926) to discover that microscopic structures in nervous tissue could be made visible by staining. Golgi's staining technique involved soaking nervous system tissue in a solution of silver salts, which made small fibers visible by turning them brownish black. These colored filaments then became visible against the background of tissue not colored by the solution. But because nervous system tissue is so densely packed, Golgi was unable to follow the tiny fibers composing the nervous system from their beginning to their end points and thus determine the fibers' structure and purpose. Adding to the problem, Golgi's staining procedure often produced uncertain results, and sections that were cut too thin or completely stained complicated efforts to follow the fibers to their terminal points.

Spanish scientist Santiago Ramón Cajal (1852–1934) dedicated himself to determining the shape and purpose of the small fibers in the central nervous system using the Golgi staining technique. After several failed attempts, Cajal said,

> I did not, however, lose faith in the method on that account. I was fully convinced that, in order to make a significant advance in the knowledge of the structure of the nerve centres, it was absolutely necessary to make use of [Golgi staining] procedures capable of showing the most delicate rootlets of the nerve fibres vigorously and selectively coloured upon a clear background.... It remained only to determine carefully the conditions of the chrome-silver reaction, and to regulate it so as to adapt it to each particular case.[1]

Given that the central nervous system of an adult mammal proved too complex to study in fine detail, Cajal came up with the idea of Golgi staining nervous tissue from younger animals and embryonic tissues. In these tissues the fibers had not yet lengthened and the myelin sheathing was not fully developed. Apparently Golgi had already tried staining younger tissue samples but gave up, thinking that such an approach would avail little in advancing our understanding of the adult nervous system. Golgi later returned to staining younger and embryonic animal nervous system tissue.

In 1888 Cajal's perseverance paid off in a manner that far surpassed his expectations. He described 1888 as "my greatest year, my year of fortune. For during this year, which rises in my memory with the rosy hues of dawn, there emerged at last those interesting discoveries so eagerly hoped and longed for."[2]

He recalled that as he labored, "new facts appeared in [his] preparations, [and] ideas boiled up and jostled each other in [his] mind."[3] Then, said he, "The new truth, laboriously sought and so elusive during two years of vain efforts, *rose up suddenly in my mind like a revelation.*"[4] That revelation enabled him to determine the general structure of the neuron and its purpose. He discovered that neurons are made up of axis cylinders (known as axons) that end in an arrangement of branch-like structures, and that these branch-like structures come into close contact with "tree-like outgrowths from the cell body"[5] (known as dendrites). Moreover, because these two structures were so close to each other, he discovered that "the cell bodies and their protoplasmic processes enter into the chain of conduction, that is to say, that they receive and propagate the nervous impulse."[6]

Cajal's inspiration unlocked the secrets of the human nervous system. His discovery revolutionized our understanding of the human body, how it responds to the environment, regulates behavior, and is affected by drugs and disease.

PENICILLIN

Throughout world history, bacterial infections have claimed innumerable lives and caused widespread fear and sorrow. Only recently has the scourge of bacterial infections been alleviated, thanks in large part to the discovery of penicillin in 1928.

The story of the discovery of penicillin began on the battlefields of France during World War I. A Scotsman serving in the medical corps, Alexander Fleming (1881–1955), was so moved by the suffering and death caused by infections that he was determined to find a cure. After the war he set up a small one-room research laboratory in London's St. Mary's Hospital, where he toiled in search of antiseptics that could kill bacteria without damaging human tissue.

One day, after returning to his lab from a summer vacation, he looked through a stack of petri dishes containing cultures of staphylococcus, a bacterium commonly found on mucous membranes. One of the petri dish cultures appeared contaminated. He was about to throw away the contents of the contaminated dish when he suddenly decided to take a second look. Upon closer examination, he saw a green mold growing in one section of the dish. In the area immediately around the mold there were very few colonies of staphylococci, and colonies that had come into contact with the mold were dead. He removed some of the mysterious mold and placed it in an environment where it could grow. When the mold matured into a green mass, he examined it under the telescope and discovered that it was *Penicillium notatum*, a fungus found on decaying organic matter like cheese and fruit.

The agent in *Penicillium*, which he called penicillin, was a powerful antibiotic. When diluted to one-hundredth of its original strength, it readily killed *staphylococci* and other deadly germs such as *pneumococci* and *streptococci*, all without harming human tissue. *This serendipitous discovery was nothing short of a miracle.* The normal course of action for finding a contaminated petri dish would have been to discard the contents, but Fleming was prompted to take a second look. It is also noteworthy that Fleming subsequently tried to produce penicillin from at least twenty other types of mold, including eight variants of the *Penicillium* fungus, but only the type of mold that ended up in his petri dish produced penicillin. Indeed, the circumstances surrounding this breakthrough make it a truly amazing discovery.

Louis Pasteur once said, "In the field of observation, chance favors the prepared."[7] This certainly was the case with Fleming. Yet I do not believe that this discovery was solely the result of chance events. Given that this chain of unlikely events led to a treatment that has saved countless lives, I am certain that Fleming's discovery is an example of the Lord working in mysterious ways to prolong life in the latter days.

ORGANIC CHEMISTRY

During the nineteenth century, chemists struggled with understanding how identical carbon atoms combined to form different compounds. This mystery was solved in a most wondrous manner by a chemist named Friedrich August Kekule (1829–1896).

Kekule was returning home late one night after discussing chemistry with his friend Hugo Muller. While riding home on an open bus through the deserted streets of London, he experienced the following:

> I fell into a reverie, and lo, the atoms were gamboling before my eyes. . . . I saw how, frequently, two smaller atoms united to form a pair; how a larger one embraced the two smaller ones; how still larger ones kept hold of three or even four of the smaller; whilst the whole kept whirling in a giddy dance. I saw how the larger ones formed a chain, dragging the smaller ones after them but only at the ends of the chain.[8]

Kekule spent much of that night drawing the shapes he saw in his dream. His sketches illustrated how carbon atoms create different substances by forming links and chains. This vision led to Kekule's theory of organic molecular structure.

Sometime after this discovery, Kekule dedicated himself to studying aromatic benzene, a hydrocarbon found in aromatic substances such as scented oils and spices. Benzene does not follow the same rules of organic molecular structuring that Kekule discovered in his first dream. After laboring for seven years to unlock the secrets of the structure of benzene that accounted for different aromatic properties, Kekule had another revelatory dream. He recalled:

> I was sitting writing at my textbook but the work did not progress; my thoughts were elsewhere. I turned my chair to the fire and dozed. Again the atoms were gamboling before my eyes. This time the smaller groups kept modestly in the background. My mental eye, rendered more acute by repeated visions of the kind, could now distinguish larger structures of manifold conformation: long rows, sometimes more closely fitted together all twining and twisting in snake-like motion. But look! What was that? One of the snakes had seized hold of its own tail, and the form whirled

mockingly before my eyes. As if by a flash of lightning I awoke; and this time also I spent the rest of the night in working out the consequences of the hypothesis.[9]

This dream led to Kekule's discovery that carbon atoms also form rings.

Kekule's dreams of carbon atoms forming chains and rings answered the question of how identical carbon-based compounds produce different substances. This discovery spawned an organic chemistry industry that today provides indispensable coal-tar products such as dyes, plastics, detergents, and drugs. His dreams also unlocked mysteries of life on earth, for all organic life "depends on the capacity of carbon atoms to form molecular chains and rings as they did in Kekule's dreams."[10]

MALARIA

Malaria is arguably one of the world's most devastating diseases. Throughout history it has killed millions of people and ravaged entire populations. The true cause of malaria remained hidden from mankind for millennia. For a while people thought that it was caused by "poisonous gasses" released from swamps and marshlands. Then during the late nineteenth century, the cause of malaria was revealed by divine inspiration to an Englishman named Ronald Ross (1857–1932).

While serving with the Medical Service in India, Ross was deeply affected by the suffering caused by malaria. The suffering he witnessed motivated him to settle the mystery surrounding the cause of the disease. His first breakthrough came when an acquaintance, who had discovered that elephantitis was caused by a parasite in mosquitoes, suggested to him that mosquitoes might also carry malaria. Ross decided to investigate this hypothesis but was unsure about how to proceed.

One evening while stationed at a military post in rural India, Ross walked outside and climbed up a pile of rocks to look over the plains. In the distance he spotted a vulture feeding on the carcass of a jackal. Then the thought came to him, Why not investigate if mosquitoes that feed on malaria-infected blood contain a parasite? Soon he was feeding mosquitoes on consenting malaria patients and then dissecting the mosquitoes under an old, worn microscope. After numerous failed attempts to locate a parasite in this way, he wrote to his wife, "I have failed in finding parasites in mosquitoes fed on malaria patients, but perhaps am not using the proper kind of mosquito."[11]

Then, "as if in answer [to his plea], some Angel of Fate . . . put into his [worker's] hand a bottle of mosquito larvae" that hatched a mosquito unlike any others Ross had been studying.[12] Ross promptly fed the new mosquitoes on a malarial patient and then set ten of them aside for later dissection. The next day he dissected two of the new mosquitoes but was unable to find the parasite. More dissections the following day showed no sign of the parasite. The situation was growing desperate, for only two of the original ten mosquitoes remained, and Ross' workers had not been able to locate any more larvae from this variety of mosquito. Nevertheless, Ross sensed that he was on the verge of a major discovery; he had a premonition that led him to believe that he was getting closer to the truth.

He described the momentous dissection of the second to last mosquito as follows: "The dissection was excellent and I went carefully through the tissues, now so familiar to me, searching every micron with the same passion and care as one would search some vast ruined palace for a little hidden treasure [but found nothing]."[13] Sitting there in front of his microscope, his attention was suddenly drawn toward the mosquito's stomach. He felt that he should examine the stomach, but resisted. He recalled, "I was tired and what was the use? I must have examined the stomachs of a thousand mosquitoes by this time."[14]

Eventually he relented and proceeded to inspect the stomach. As he analyzed its contents, he declared, "The Angel of Fate fortunately laid his hand on my head," for no sooner had he examined the contents of the stomach when, he said, "I saw a clear and almost perfect circular outline before me of about 12 [microns] in diameter."[15] In several of these cells he found clusters of small, black granules. It was the malaria parasite he had labored so long and hard to find.

Ross discovered that the malaria parasite is carried by the *Anopheles Stephensi* variety of mosquito, that the parasite lives in the mosquito's digestive system, and that it infects humans through the mosquito's saliva. This discovery led to public health measures that have curbed the spread of the disease and saved countless lives. Ross would not take all the credit for his discovery, though. On that fateful night in which he found the parasite, he penned the following poem acknowledging the Lord's assistance in this great discovery:

> This day designing God
>   Hath put into my hand
> A wondrous thing. And God
>   Be praised. At His command,

> I have found thy secret deeds
>   Oh million-murdering Death.
>
> I know that this little thing
>   A million men will save—
> Oh death where is thy sting?
>   Thy victory oh grave?[16]

THE PERIODIC TABLE

In September 1860, an international conference was held in Karlsruhe, Germany, to resolve the issue of how chemists should measure the weights of elements. Two approaches were being considered, the atomic weight method and the equivalent weight method. In attendance was an obscure twenty-six-year-old chemist from Siberia, Dmitri Mendeleyev. During the conference, Mendeleyev was greatly influenced by a speech given by atomic weight proponent Stanislao Cannizzaro. Mendeleyev was impressed by the energy of Cannizzaro's speech and by how Cannizzaro "admitted of no compromise and seemed to advocate truth itself."[17]

During the nineteenth century, chemistry was still in its infancy, struggling to evolve into a mature scientific enterprise. Unlike physics, which had long since settled on Newtonian mechanics as its working paradigm, chemistry lacked a unifying theory or principle. Moreover, mysticism apparently continued to influence the views of some nineteenth century chemists. Even Mendeleyev's senior colleague at the University of St. Petersburg, the prominent chemist Aleksandr Butlerov, had an tendency to integrate scientific and mystical explanations. Mendeleyev bravely debunked some of Butlerov's more outlandish claims.

Mendeleyev's yearning for truth may have been influenced by his dying mother's advice. She urged him to "refrain from illusions, insist on works and not on words," and to "patiently seek divine and scientific truth."[18] These words impressed Mendeleyev so much that he wrote them in a scientific treatise dedicated to his mother's memory.

In 1869 he completed his first volume of *The Principles of Chemistry*, a book heralded as one of the finest late nineteenth century works on chemistry. Shortly after beginning work on the second volume, he encountered the problem of how best to organize the elements. He brought his twenty years of study and experience to bear on this problem. On February 14, 1869, he began searching in earnest for an underlying principle by which the sixty-three known elements could be arranged. He tried organizing the elements by their atomic weights, from lightest to heaviest, but doing so produced no clear pattern. He labored tirelessly on this problem for the next three days, alone in his study. Author Paul Strathern described the scene as follows:

> Mendeleyev sat at his littered desk amidst the clutter of his study: a wild-haired, gnome-like figure obsessively combing the fingers of his left hand through the points of his long straggly beard. From the dimness above his head the portraits of Galileo, Descartes, Newton, and Faraday stared down as his pen scratched away amidst the encroaching disorder of spilling papers, books, and obscure mechanical devices.[19]
>
> A few days later, on the afternoon of seventeenth, his friend Inostrantzev paid him a visit. Inostrantzev found Mendeleyev in a frustrated state. Mendeleyev felt that he was getting closer to a solution that would allow him to organize the elements in a comprehensive fashion, but lacked a satisfactory principle of organization. "It's all formed in my head," he exclaimed, "but I can't express it."[20]

Shortly after Inostrantzev's visit, Mendeleyev took sixty-three cards and wrote down the known elements and their properties. He shuffled these around as he searched for an underlying pattern. As he grouped the cards by their chemical properties, he noticed that the values of the atomic weights tended to fall into alignment as well. It was then that he fell asleep from exhaustion and began to dream.

Mendeleyev recalled, "*I saw in a dream a table where all the elements fell into place as required.* Awakening I immediately wrote it down on a piece of paper."[21] The image that emerged in his sketch was a table that grouped the elements by their atomic weights and chemical properties. The organization that he saw in his dream revealed that "the properties of elements 'were periodic functions of their atomic weights;'"[22] thus the name, the Periodic Table.

Mendeleyev soon discovered inconsistencies between his periodic table and commonly accepted atomic weights of known elements. For instance, tellurium was thought to have an atomic weight of 128, yet Mendeleyev's table indicated that its atomic weight was somewhere between 123 and 126. Also, gold was thought to have an atomic weight of 196.2, but Mendeleyev's table indicated that gold had a greater atomic weight than platinum's 196.7. Notwithstanding stiff skepticism from other chemists, these corrections were eventually validated.

In addition, after Mendeleyev assembled the know elements according to the organization revealed in his dream, he noticed that there were vacant positions. This allowed him to identify "missing" elements such as eka-boron (now known as scandium), eka-aluminum (now known as gallium), and eka-silicon (now known as germanium) years before they were discovered. He was also able to predict the atomic weight and specific gravity of these "missing" elements with amazing accuracy. Since then, the table has also aided in the discovery of other elements.

The underlying principle of organization revealed in Mendeleyev's dream provided a much-needed foundation for modern chemistry. His discovery also advanced our understanding of an atomic element's ability to combine with other elements (i.e., valence), and provided valuable information regarding chemical properties that arise when elements are combined to form molecular compounds. These important contributions revolutionized chemistry and played a major role in shaping the world in which we live.

LOWER MORTALITY FROM DISEASE

The previous accounts are evidence of the Lord blessing humanity with knowledge in the last days. There is also an indirect yet equally impressive way that the Light of the Lord has blessed humanity. This blessing is a significant decrease in deaths caused by infectious disease shortly after the Restoration of the gospel and decades before the advent of twentieth century medications that cure and prevent disease.

Research on deaths rates shows that there was a sharp decrease in the number of Americans dying from infectious diseases during the late nineteenth and early twentieth century. For example, during the late nineteenth and early twentieth centuries, deaths from measles dropped dramatically before the creation of a vaccine in the early 1960s. Deaths from pneumonia dropped significantly before the creation of sulphonomide during the 1930s. Deaths from tuberculosis declined sharply before the creation of izoniazid during the 1930s. Deaths from typhoid dropped dramatically before the creation of chloramphenicol during the 1940s. Deaths from diphtheria declined before the creation of toxoid during the 1930s. And deaths from polio declined gradually before the creation of a vaccine by Jonas Salk during the 1950s.[23]

What are the reasons for these decreases in mortality? Emphasis on better health care practices, nutrition, and improved living conditions likely played a role. But as Leonard Sagan points out in his 1987 book *The Health of Nations: True Causes of Sickness and Well-Being*, these factors alone do not account for the decreased deaths. Sagan posited, "More important in explaining the decline in death worldwide is the rise of hope and the decline in despair and hopelessness."[24]

What was the source of this "rise in hope" during the nineteenth and twentieth centuries? I believe that it was the outpouring of the Spirit of the Lord in the latter days and the concomitant Restoration of the gospel of Jesus Christ. Ezra Taft Benson pointed out that "the restored gospel of Jesus Christ *brings hope to all the world.*"[25] Apostolic authority, priesthood power, saving ordinances of the gospel, and the Spirit of the Lord bring blessings of hope to all people, regardless of whether they are members of the Church or not. Once again, we see that the presence of the enlightening power of the Spirit of the Lord and the presence of the fulness of the gospel are essential to the well-being of mankind.

The words of Mormon appropriately summarize these and other secular blessings. He wrote, "Behold, are not the things that God hath wrought marvelous in our eyes? Yea, who can comprehend the marvelous works of God?" These marvelous discoveries manifest His love for humanity, the power of His Spirit, and His desire to bless humanity with knowledge and truth.

Notes
1. *The Faber Book of Science.* Edited by John Carey (Boston, MA: Faber and Faber, 1995), 253.
2. Ibid., 254.
3. Ibid., 257.
4. Ibid., 254; emphasis added.
5. Ibid.
6. Ibid., 255.
7. Brian L. Silver, *The Ascent of Science* (New York: Oxford University Press, 1998), 87.
8. *August Kekule and the Birth of the Structural Theory of Organic Chemistry in 1858.* Translated by O. Theodor Benfey in *Journal of Chemical Education,* 1958, vol. 35., 22.
9. Ibid.
10. *The Faber Book of Science.* Edited by John Carey (Boston, MA: Faber and Faber, 1995), 138.
11. Ibid., 206.
12. Ibid.
13. Ibid., 208.
14. Ibid.
15. Ibid.
16. Ibid., 209.
17. Paul Strathern, *Mendeleyev's Dream: The Quest for the Elements* (New York: Thomas Dunne, 2000), 273.

18. Ibid., 264–65.
19. Ibid., 279.
20. Ibid., 283.
21. Ibid., 286; emphasis added.
22. *The Faber Book of Science*. Edited by John Carey (Boston, MA: Faber and Faber, 1995), 155.
23. J. B. McKinlay and S. M. McKinlay *Medical Measures and the Decline of Mortality, in The Sociology of Health and Illness: Critical Perspectives*. Edited by Conrad, P. and Kern, R. (New York: St. Martin's Press, 1981), 25.
24. Leonard Sagan, *The Health of Nations: True Causes of Sickness and Well-Being* (New York: Basic Books), 1987, 184.
25. Ezra Taft Benson, *God, Family, Country: Our Three Great Loyalties* (Salt Lake City: Deseret Book, 1974), 249; emphasis added.

SEVEN

SKEPTICS AND BELIEVERS

The absence of chance and the serving of ends are found in the works of nature especially.

—ARISTOTLE

THEISM

Theism is believing that a supreme being created this world for a purpose and that He is actively involved in His creations. Our Latter-day Saint belief system is theistic. We believe that the world was created for the purpose of "bring[ing] to pass the immortality and eternal life of man" (Moses 1:39). We also believe that God is personally interested in the progress of His children and that He hears and answers prayers. Mosiah summed up our theism by declaring that God "has created you from the beginning, and is preserving you from day to day, by lending you breath, that ye may live and move and do according to your own will, and even supporting you from one moment to another" (Mosiah 2:21).

Theism also claims that God is immanent in the world, which means that He exists in and through His creations. This view is consistent with LDS doctrine in terms of the Light of Christ being immanent. Doctrine and Covenants 88:12–13 teaches that the Light of Christ "proceedeth forth from the presence of God to fill the immensity of space." This Light "is in all things, . . . giveth life to all things, . . . [and] is the law by which all things are governed, even the power of God who sitteth upon his throne, who is in the bosom of eternity, who is in the midst of all things."

Doctrine and Covenants 88:41 similarly declares that He is "in all things and is through all things; and all things are by him, and of him."

Theism also claims that the creator transcends the world in the sense of being superior to and separate from the world. This view is consistent with our belief that "the heavens is a place where God dwells and all his holy angels" and that "he looketh down upon all the children of men, and he knows all the thoughts and intents of the heart" (Alma 18:30, 32). His transcendence is also suggested by the familiar scripture: "For my thoughts are not your thoughts, neither are your ways my ways, saith the Lord. For as the heavens are higher than the earth, so are my ways higher than your ways, and my thoughts than your thoughts" (Isaiah 55:8–9).

Many prominent scholars who lived during the second illumination were theists. For instance, Galileo believed that God actively communicates wisdom to mankind by providing individuals with inspiration.[1] Descartes believed that God preserves the world with the same degree of involvement as when He created the world.[2] And Isaac Newton believed that continued divine influence is necessary to maintain life and order in the world.[3]

In my opinion, one of the reasons these scholars were theists was they felt the enlightening power of the Spirit of the Lord in their work. The Apostle Paul taught that the deep things of God are revealed through the Holy Spirit (1 Corinthinas 2:10–11). Through their association with the Spirit, these men came to know that God created this world for a divine purpose and that He is personally involved in its affairs. If they had been intellectually arrogant and close minded, they would not have received this inspiration. Paul stated that "the natural man receiveth not the things of the Spirit of God: for they are foolishness unto him: neither can he know them, because they are spiritually discerned" (1 Corinthians 2:14).

THE ENLIGHTENMENT

Unfortunately, theism began to wane as Western Europe entered into a new era known as the Enlightenment (circa early 1700s). The Enlightenment, which lasted for most of the eighteenth century, gets its name from the fact that people who lived then believed that they were living in enlightened times. They believed that they were living in an age that was far more civilized and advanced than previous generations. With respect to scholarly endeavors, they were correct; theirs was a time of prosperity unmatched by previous generations. However, their so-called Enlightenment was un-enlightening in a spiritual sense. It produced a spiritual malaise in secular scholarship that continues to this day.

Theism began to wane when Enlightenment scholars downplayed the role of God in the universe. Efforts to minimize the role of deity were largely spearheaded by French thinkers known as the philosophes, men like Diderot (1713–1784), Voltaire (1694–1778), and Montesquieu (1689–1755). The philosophes were influential writers and publicists who rewrote, in the vernacular, abstruse scholarly works that the general population had been unable to understand. In their writings the philosophes minimized the influence of a higher power and elevated human reason to a position of prominence.[4] As a result, many Enlightenment scholars grew prideful of their intellectual and scientific achievements and denied the influence of the Almighty.

One can only imagine how theists of the second illumination would have felt about these changes. Historian Brian Silver gives us some idea of how Newton might have felt. He wrote, "Newton neither foresaw nor intended any of this. He was not the John the Baptist of [i.e., the one who prepared the way for] the Enlightenment, and he would not have been at home with its ideals."[5] I am certain that the same could be said for other inspired scholars like Robert Boyle, Descartes, and Galileo.

DEISM

The decline of theism during the Enlightenment was accompanied by the rise of deism. Deism, which is still popular today, was described by Bruce R. McConkie as "the partial acceptance of God." He observed that "deists profess to believe in him as the Creator of the world . . . but they reject the idea that he rules over or guides men during the interval between the creation and the judgment."[6] In other words, deists believe that the Lord is a disinterested creator, whose only involvement with humanity occurred during the Creation. They claim that after the Creation, the creator left the world to run on its own according to the laws of nature. In a way, the deist god is like a watchmaker who, after building a watch, has no further involvement in its operation.

Let's take a look at a few concerns with deism. To begin with, expressions of faith are a low priority. If, as the deists suppose, the creator is not paying attention, then there's no reason to pray, attend church, and keep the Sabbath Day holy. We may ask, if deists are not willing to express their faith, then why do they choose to believe in God? One answer is that, although they believe in a creator, acknowledging His existence is of limited importance because He does not influence events on earth. What is important is discovering, through science, the laws of nature that influence the world. Science is thus recognized as the only reliable source of knowledge about the natural world and the key to humanity's happiness and survival. Believing that science is the only reliable source of truth and that it holds the answers to all of life's mysteries is called scientism.

What many theists do not realize is that it is impossible to accept God and scientism. The Savior taught, "No man can serve two masters; for either he will hate the one and love the other, or else he will hold to the one and despise the other. Ye cannot serve God and Mammon" (3 Nephi 13:24). Deists who embrace scientism are clearly worshiping scientific mammon.

Another problem is that the notion of an uncommunicative and unknowable creator has led to some rather confusing descriptions of God. Consider this unusual description of the creator by a scholarly deist:

> [The creator is] the ground and source of our sense of wonderment, of power, of powerlessness, of light, of dark, of meaning, and of bafflement.... It is the God of mystics of all cultures and creeds. We stand on the shore of knowledge and look out into the sea of mystery and speak his name. His name eludes all creeds and all theories of science. He is indeed the "dread essence beyond logic."[7]

Other deists equate God with nature, a belief known as pantheism. Benedict Spinoza, a seventeenth century scholar and deist, promulgated the view that the creator *is* nature. He believed that God is the structure of the cosmic order, which operates according to universal laws of nature without divine purpose. Thus, "Spinoza's God . . . [can]not be spoken to, [does] not respond if prayed to, [and is] very much in every particle of the universe."[8] After reading these descriptions, is it any wonder that deists view the creator as a detached and unknowable entity? How could anyone commune with such a being?

In order to truly understand nature, we must know the creator. Knowing that this world was created for the purpose of "bring[ing] to pass the immortality and eternal life of man" (Moses 1:29) engenders respect for the Lord and enables us to find deeper meaning in His creations. To illustrate, in *Skeptics and True Believers*, deist and science author Chet Raymo expresses amazement over a Hubble Deep Field photograph that shows over 1,500 galaxies in one photo of the night sky. His wonderment stems from being able to witness those galactic structures, knowing what they are, and having some knowledge of the natural processes by which they were formed.

I am similarly amazed whenever I look at photographs of the Hubble Deep Field, but there is much more to my amazement. I feel immense respect for the greatness and power of God. I am amazed that, notwithstanding His greatness and power, He is merciful, just, and loving (Alma 42:15; 26:37). Despite His innumerable creations, He is keenly aware of each one of us and concerned for our well-being (Matthew 7:11; Mosiah 4:21). I am also amazed that Heavenly Father wants to share all He has with us, and have us participate in similar creations (D&C 98:18; 132:19–20). Indeed, knowing that God's purpose is to bring to pass the immortality and eternal life of man (Moses 1:39) has enriched my understanding of our world and the cosmos.

Perhaps the most troubling aspect of deism is that the belief in divine non-involvement prevents people from petitioning God for assistance and, more importantly, from giving thanks to Him for their discoveries. Joseph F. Smith called this the sin of ingratitude. He said,

> In all the great modern discoveries in science, in the arts, in mechanics, and in all material advancement of the age, the world says, "We have done it." The individual says, "I have done it," and he gives no honor or credit to God. Now, I read in the revelations through Joseph Smith, the prophet, that because of this, God is not pleased with the inhabitants of the earth but is angry with them because they will not acknowledge his hand in all things.[9]

The Lord has instructed us to give thanks for every blessing we receive (D&C 46:32). We can only do this if we recognize Him as a continual source of truth and light in the world.

AGNOSTICISM AND ATHEISM

As the Enlightenment progressed and deists relegated the role of deity to the Creation, it was all too easy for nonbelievers to step in and assert their skeptical beliefs. These circumstances led to the rise of agnosticism and atheism during the eighteenth and nineteenth centuries. During this time, reference to deity in scientific works was conspicuously absent, compared to the seventeenth century writings. Two popular theories that illustrate this change are Pierre-Simon Laplace's (1749–1827) nebular theory of the creation of the solar system and Charles Darwin's theory of evolution.

Laplace proposed that the solar system was created from a nebular gas that condensed and coalesced into planets as it swirled around the sun.

At the time, this theory was considered iconoclastic because it failed to mention the role of deity in the process. Noting the theory's failure to mention deity, Napoleon Bonaparte (1769–1821) commented to Laplace, "Monsieur Laplace, they tell me you have written this large book on the system of the universe and have never even mentioned its Creator." Laplace replied, "Sire, I had no need for that hypothesis."[10]

Deity was also conspicuously absent from Charles Darwin's theory of evolution. On October 11, 1859, Darwin addressed the issue of divine non-involvement in a letter to Charles Lyell. He wrote: "I would give absolutely nothing for [the] theory of natural selection if it required miraculous additions at any one stage of descent,"[11] suggesting that he rejected the notion of deity playing a role in the origin of species.

Enlightenment agnosticism and atheism continue to dominate science today, as does the legacy of leaving out or denying deity in scholarship. Given their influence on modern science, these two belief systems deserve closer examination.

Agnosticism is indecision about whether God lives. Many agnostics were taught at a young age to believe in a supreme being but called their faith into question after learning scientific principles in school. Ezra Taft Benson acknowledged this problem when he said, "Students at universities are sometimes so filled with the doctrines of the world they begin to question the doctrines of the [Lord's] gospel."[12] Rather than completely deny the religious lessons of their youth, agnostics put their faith on hold pending conclusive evidence confirming the existence of God.

However, by provisionally denying the existence of God, agnostics are not much different from atheists, who categorically deny the existence of God. In particular, agnostic and atheist scholars both believe that everything that happens in the world is simply matter in motion, that world is continually driven to and fro without purpose according to the "blind" laws of nature.

What is surprising is that agnostics and atheists adhere to their doctrine of disbelief despite ample evidence for the existence of God. As Alma told the antichrist, "All things denote there is a God, yea even the earth and all things that are upon the face of it, yea, and its motion, yea, and also all the planets which move in their regular form do witness that there is a supreme creator" (Alma 30:44). Yet Korihor chose not to believe. People's refusal to acknowledge evidence for the existence of God troubled Nephi who said, "I am left to mourn because of the unbelief . . . ignorance, and stiffneckedness of men; for they will not search knowledge, nor understand great knowledge, when it is given unto them in plainness, even as plain as word can be" (2 Nephi 32:7).

Science and religion writer Phil Johnson suggests that atheists choose not to see the evidence for the existence of a supreme being because "they don't want to find evidence for what they think of as an 'interfering' God, meaning a God who does not leave everything to law and chance."[13] He continued, "Hence they will refuse to see evidence of design that is staring them in the face until they are reassured that the designer is something whose existence they are willing to recognize."[14]

Scholars who refuse to see the abundant evidence for the existence of God have been "blinded by the craftiness of men" (D&C 76:75). Self-proclaimed agnostic and renowned scientist Stephen J. Gould alluded to this blinding power when he admitted that an atheistic theory once "beguiled [him] with its unifying power when [he] was a graduate student in the mid-1960s, [but] since then [he had] been watching it slowly unravel."[15]

How, then, do atheists interpret the evidence for the existence of deity mentioned by Alma? In short, they believe that the earth and all life came into existence by basic laws of nature and random processes. Problems with this hypothesis abound. To begin with, the very idea of natural law, which they readily accept, denotes a lawgiver. Moreover, without a divine purpose to guide their application, laws of nature are incapable of producing complex living systems. But, the atheists argue, complex systems may evolve through random processes. This typical response contains a contradiction. Chance events are incapable of producing complex systems *because they are random*. As science author Brain Silver pointed out, randomization demonstrates "no preferred direction for a system to evolve."[16]

The unlikelihood of complex systems evolving from chance is illustrated by the second law of thermodynamics. According to this law, matter in the universe is continually moving towards high entropy, or a disordered state. Yet within the universe there is low entropy in the form of complex life. Somehow "against the overall tendency of the universe to go to ever higher entropy," life has "built itself out of a high-entropy (disordered) world and it maintains its low-entropy (order)."[17]

The second law of thermodynamics does not preclude order (low entropy) from coming about by chance; it suggests that spontaneous, undirected creation of complex biological systems is very unlikely. Without intelligent design to organize matter, elements exist in a disordered state. This was the case with the physical elements of our earth. Before the Lord assembled the primordial material that composes our world, it was "matter unorganized."

A typical atheistic response to this intelligent design argument is that complex living systems can evolve by chance, given millions of years. Consider, for example, the following statement by skeptic and chemist Peter Atkins. He declared, "I think it is much simpler to say that universes just tumble into being (whatever the means), and if one of them happens to have the right mix, then life will take hold in it. This is intrinsically much simpler than a designed universe."[18] This counterargument illustrates what I call the belief in the magic of high numbers. It is the belief that as the number of trials needed for bringing about a random event increases over time, the probability of that event occurring at any one moment in time increases.

The belief in the magic of high numbers is a fallacy. Consider that the more times you toss a coin, the greater your opportunities for getting heads; but innumerable tosses do not change the fact that with each toss of the coin, the probability of getting heads is always 50 percent. Similarly, given that the probability of matter randomly organizing into a complex system is extremely unlikely, regardless of the number of "attempts" to organize, the probability that matter will organize into a complex system at any one moment in time remains extremely unlikely. No matter how many millions of years have passed and how many more years lay ahead, at any given point in time, the improbable remains improbable.

From a purely rational standpoint, then, I agree with philosophy professor Roger Trigg who said, "It's much simpler to believe in God who created the one universe, rather than saying there are an enormous number [of universes] and we just happen to be in the one that's come up in this way."[19]

I find it surprising that so many cling to atheism, given its pessimistic outlook. Atheism assumes, of course, that when life is over, it is over. On a few occasions I have tried to imagine this scenario. This cognitive exercise usually results in feelings of despair. The despair always dispels as I reflect on my testimony of Heavenly Father, the Savior, and the plan of redemption. Knowing that God lives and that we are eternal beings brings a sense of joy and hope that's unmatched by secular aphorisms such as "We live on in the genes of our children" and "The indomitable human will lives on."

Atheists may experience similar feelings of joy and hope if they exercise faith. Alma wrote:

> If ye give place, that a seed [of faith] may be planted in your heart, behold, if it be a true seed, or a good seed, if ye do not cast it out by your unbelief, that ye will resist the Spirit of the Lord, behold, it will begin to swell within your breasts; and when you feel these swelling motions, ye will begin to say within yourselves—It must needs be that this is a good seed, or that the word is good, for it beginneth to enlarge my soul; yea, it beginneth to enlighten my understanding. (Alma 32:28)

Blaise Pascal, the eminent eighteenth century French mathematician, unknowingly performed Alma's faith experiment. One day while riding a stagecoach, his horses began galloping uncontrollably toward a bridge. Suddenly, for no apparent reason, the traces broke free and the carriage came to a stop as the horses plummeted over the edge. This event prompted Pascal to more fully exercise faith in God. It was a decision that changed his life.

For scholars requiring a more rational justification for believing in God, Pascal devised what has come to be known as Pascal's wager. According to Pascal's wager, if there is no God, then the rewards of believing and not believing are the same. If, on the other hand, God lives, then the rewards of believing and accepting Him are much greater than the rewards of choosing to not believe.

I think King Mosiah summed it up best when he said, "Believe in God; believe that he is, and that he created all things, both in heaven and in earth; believe that he has all wisdom, and all power, both in heaven and in earth" (Mosiah 4:9). The prophet Alma similarly counseled, "Even if ye can no more than desire to believe, let this desire work within you" (Alma 32:27). The blessings of exercising faith are clear. The Lord has promised, "if any man hear my voice, and open the door, I will come in to him, and will sup with him, and he with me" (Revelation 3:20). But until "the wise, and the learned . . . come down in the depths of humility" and exercise a particle of faith, "he will not open unto them" (2 Nephi 9:42).

SECULARISM

The rise of secularization during the Enlightenment raised concerns among prominent individuals of the time. Noting the growing influence of secularism in science, Queen Victoria (1837–1901) cautioned: "Science is greatly to be admired and encouraged, but if it is to take the place of our Creator, and if philosophers and students try to explain everything and to disbelieve whatever they cannot prove, I call it a great evil instead of a great blessing."[20] British Prime Minister William Gladstone (1809–1898) similarly cautioned, "Let the scientific men stick to their science and leave philosophy and religion to the poets, philosophers and theologians."[21]

As secularism spread into other areas of society, it began to replace traditional Christian values. Historians Palmer and Colton described secularism's impact on society's Christian values as follows:

> The older Christian view seemed no longer to be necessary. Thinkers provided theories of society, world history, human destiny, and of the nature of good and evil in which Christian explanations had no part. Old Christian virtues, such as humility, chastity, or the patient bearing of pain and sorrow, ceased to be regarded as significant (except in some ways by Rousseau). Christian love was transformed and secularized into humanitarian good will.[22]

The secularization of society that began during the Enlightenment continues today. Modern-day proponents of secularism now rally under a new banner called secular humanism. Secular humanists have organized into groups with the purpose of removing God from all of our public institutions. They deny supernatural authority, reject divine guidance, place all their trust in human reason, and teach that humanity's problems can be solved through scientific discovery and technology.

Some secular humanists have even attempted to replace religion with secularist values. For example, Nobel Laureate chemist Wilhelm Ostwald suggested that "scientific knowledge could replace religion as a foundation for morality and happiness," and insisted that "science *is* the god of the modern world"[23] (emphasis added). Richard Dawkins, the scholar who likened religious belief to a computer virus, stated that "most people who are influenced by religion are, I believe, impoverished rather than enriched by it. Without it, they would have a vision of the Universe, of life, and of their place in it which is bigger, more dignified, more uplifting."[24] The equally pessimistic Peter Atkins stated, "I don't think there is any question that science cannot tackle. And I think that, as it tackles them, it gives people answers that are much more reliable, much more plausible, than the obscure arguments religion provides."[25]

Secular humanists fail to realize that secular progress is inseparably connected to the enlightening power of the Spirit of the Lord. Without the enlightening power of the Lord's Spirit, humanity would flounder in ignorance and darkness. In addition, by attempting to replace Christian values with secular values, the humanists are turning the Lord's blessings of illumination into a curse. Scholars who apply the blessings of illumination in a manner contrary to the will of the Lord by attacking gospel beliefs often fall into the trap ever learning without coming to the knowledge of truth. Though they know it not, they have become so "puffed up [in] their learning, and their wisdom" (2 Nephi 9:42) that they are no longer able to perceive the things of God. They are forfeiting eternal happiness for worldly learning. As Nephi declared, "To be learned is good *if* [one] hearken[s] unto the counsels of God" (2 Nephi 9:30; emphasis added).

It is important for Latter-day Saints to recognize that we are in a struggle against forces of evil masquerading as human progress. As defenders of truth and righteousness, Latter-day Saints need to be aware of the growing influence of secular humanism and counter its effects on our families and communities. As we do so, we may take comfort in knowing that because the fulness of the gospel of Jesus Christ will never again be taken from the earth, truth and righteousness will continue to grow until the Savior comes.

Notes

1. *Discoveries and Opinions of Galileo.* Translated with an introduction and notes by Stillman Drake (New York: Doubleday Inc., 1957), 190–91.
2. René Descartes, *Discourse on Method.* 3d ed. Translated by Donald Cress (Indianapolis, Indiana: Hackett Publishing, 1998), 24–25.
3. James Gleick, *Isaac Newton* (New York: Pantheon Books, 2003), 103.
4. For example, see R.R. Palmer and Joel Colton, *A History of the Modern World.* 3d ed. (New York: Alfred A. Knopf, 1965), 290–300.
5. Brian Silver, *The Ascent of Science* (New York: Oxford University Press, 1998), 57.
6. Bruce R. McConkie, *Mormon Doctrine.* 2d ed. (Salt Lake City: Bookcraft, 1966), 189.
7. Chet Raymo, *Skeptics and True Believers* (New York: MJF Books, 1998), 214.
8. Antonio Damasio, *Looking for Spinoza: Joy Sorrow, and the Feeling Brain* (New York: Harcourt, 2003), 22.
9. Joseph F. Smith, *Gospel Doctrine: Selections from the Sermons and Writings of Joseph F. Smith.* Compiled by John A. Widtsoe, (Salt Lake City: Deseret Book Co., 1939), 270.
10. Ian Barbour, *Religion and Science: Historical and Contemporary Issues* (San Francisco: Harper Collins, 1997), 35.
11. See Randal Keynes, *Darwin, His Daughter, and Human Evolution,* (New York: Riverhead Books, 2001), 256.
12. Ezra Taft Benson, *The Teachings of Ezra Taft Benson* (Salt Lake City: Bookcraft, 1988), 302.
13. Phillip E. Johnson, *The Wedge of Truth: Splitting the Foundations of Naturalism* (Downers Grove, IL: Inter Varsity Press, 2000), 129.
14. Ibid.

15. Ibid., 71.
16. Brian Silver, *The Ascent of Science* (New York: Oxford University Press, 1998), 219.
17. Ibid., 229.
18. Russell Stannard, *Science and Wonders: Conversations about Science and Belief* (Boston, MA: Faber and Faber, 1996), 25.
19. Ibid., 30.
20. Brian Silver, *The Ascent of Science* (New York: Oxford University Press, 1998), 130.
21. Ibid.
22. R. R. Palmer and J. Colton, *A History of the Modern World* (New York: Alfred A. Knopf, 1965), 299.
23. Richard Olsen, *Physics, In Science and Religion: A Historical Introduction.* Edited by Gary B. Ferngren (Baltimore, Maryland: John Hopkins University Press, 2003), 305.
24. Russell Stannard, *Science and Wonders: Conversations about Science and Belief* (Boston, MA: Faber and Faber, 1996), 165.
25. Ibid., 167.

EIGHT
GOSPEL TRUTH AND SCIENCE

Everyone who is seriously involved in the pursuit of science becomes convinced that a spirit is manifest in the laws of the universe—a spirit vastly superior to that of man, and one in the face of which we with our modest powers must feel humble. In this way the pursuit of science leads to a religious feeling of a special sort.

—Albert Einstein

The word *science* brings many thoughts and images to people's minds. Science may be considered a body of knowledge, a community of researchers, and a social enterprise. Certainly, science is all of these. But when people think of science, what often comes to mind is that science is an activity—it is what scientists do. The activity most often associated with science is the scientific method. The scientific method refers to processes of formulating and testing research questions according to rules and procedures recognized by the scientific community.

An important dimension of scientific activity is theorizing. Theorizing is an imaginative process involving quiet reflection. Its role in science is under-recognized because it is not considered a formal testing and data-gathering procedure, yet it is equally important in the discovery process. According to the influential philosopher of science Karl Popper, theorizing "dominates the experimental work from its initial planning up to the finishing touches in the laboratory."[1]

Quiet study and reflection is a well-spring of scientific discovery. Thoughts and ideas often come to people who prepare themselves to receive inspiration through the Light of the Lord. Brigham Young taught that "there is nothing known except by the revelation of the Lord Jesus Christ, whether in theology, science, or art."[2] Commenting on this revelatory process, he added,

> There are *men of talent, of thought, of reflection, and knowledge* in all cunning mechanism; they are expert in that, though they do not know from whence they receive their intelligence. The Spirit of the Lord has not yet entirely done striving with the people, offering them knowledge and intelligence; consequently, *it reveals unto them, instructs them, teaches them, and guides them*. . . . though they know it not.[3]

Thus the Spirit reveals, instructs, and teaches those who seek to magnify their understanding through diligent study and reflection. We need to be sufficiently calm to receive its promptings, and sufficiently prepared to understand its instruction. Consider, for example, Isaac Newton's discovery of the universal law of gravity, and Albert Einstein's discovery of the general theory of relativity. When Newton was asked how he discovered the universal law of gravity, he replied, "By thinking on it continually."[4] Also, describing the moment when he discovered how to combine the theory of special relativity and the law of gravity, Einstein replied, "All of the sudden, a thought occurred to me: if a person falls freely, he will not feel his own weight."[5] This thought led to the theory of general relativity.

The source of inspiration for these two discoveries was definitely the Spirit of the Lord. D&C 11:13 teaches that when the Lord's Spirit inspires people, it *enlightens* their minds. Newton experienced this enlightening as he pondered gravity. He recalled that as he theorized, the universal law of gravity "opened slowly, by little and little, into a *full and clear light*"[6] (emphasis added). In addition, D&C 11:13 tells us that when people are enlightened by the Spirit of the Lord, their souls are filled with joy. Einstein experienced this joy when he was enlightened by the Spirit of the Lord. He described his sudden insight into general relativity as "the happiest thought of [his] life."[7]

The Spirit of the Lord also inspires those who diligently pursue technological breakthroughs. Consider that the world would not have experienced rapid scientific progress over the last few centuries were it not for earlier advancements in paper, printing, lens, and oil lamp technology. Oil lamps allowed people to read comfortably into the night, books allowed people to preserve and share ideas, and eyeglasses allowed people with poor eyesight to study literature. These technologies, which were "inspired of God for his honor and glory and the accomplishment of his work in the earth,"[8] played a major role in creating an environment in which scientific progress and the restored gospel could flourish.

THE GOAL OF SCIENCE

In 1611, the eminent scientist Galileo Galilei left his home in Florence, Italy, to visit Rome. His arrival in Rome was a jubilant affair, owing to his recently published book, *The Starry Messenger*, which revealed many fascinating astronomical discoveries. The first five hundred copies of the book sold out quickly, and orders for more copies were coming in from all across Europe. The popularity of his telescope added to his fame. Galileo's telescope was so popular that when a friend received a package from him, the friend's neighbors insisted that he open it instantly, thinking it might contain a Galilean telescope. After finding Galileo's book inside, the neighbors implored him to read aloud Galileo's description of the Medicean Stars (Jupiter's moons) that very night.[9]

Galileo's discoveries were exciting and welcome news for many, but for others they posed a threat to the centuries-old Aristotelian view of the universe. Galileo encountered stiff opposition from academicians and theologians faithful to the Aristotelian view. The debate further intensified when Galileo published his letters on sunspots in 1613. His observations of dark blemishes moving across the surface of the sun were considered heresy in light of the church-sponsored Aristotelian doctrine that the heavens were unchanging.

Interestingly, both sides of the debate adopted a sort of "what would Aristotle do?" approach to strengthen their positions. Aristotelian supporters claimed that if Aristotle were alive and aware of Galileo's discoveries, he would not change his mind on the unalterability of the heavens. The following statement by an ardent Aristotelian supporter, Lodovico delle Colombe, illustrates their unwavering devotion. Colombe said, "I should like to become in this regard an anti-Galileo out of respect to [Aristotle] that great leader of academies."[10] Galileo, on the other hand, suggested that even Aristotle would abandon his belief in the unalterability of the heavens if he had the benefit of 2,000 more years of observation and a telescope.[11]

Galileo accused his detractors of attempting "to learn from Aristotle that which he neither knew nor could find out, rather than consult [their] own senses and reason."[12] That is to say, rather than examine the astronomical data for themselves, the scholastics sought knowledge from an ancient philosopher who was not acquainted with seventeenth century issues. Repeated claims that Galileo's discoveries were mere illusions led Galileo to conclude that his detractors were "philosophizing without any desire to learn the truth and the causes of things."[13] That scientists and philosophers should have no regard for the truth when it stared them in the face, astonished Galileo. As one who desired "nothing but to win a knowledge of the truth,"[14] Galileo clearly believed that the purpose of science was to reveal truth.

Today there are belief systems which reject the idea that the purpose of science is to reveal truth. One such position is antirealism, which is essentially agnostic with regard to the truth of theories and entities that cannot be directly observed. It takes the perspective, "If we can't see, touch, or smell it, then we must assume that it's not real." To deny truth to scientific theories and unobservable entities begs the question, "Then what are our theories for?" The anti-realists would likely respond, "They serve a practical purpose." They claim that the sole purpose of science is to generate knowledge that has practical applications. An analogous viewpoint is found in agnostic attitudes toward God. Agnostics believe that it is impossible to know whether God exists because we cannot observe Him. If asked, "Then why do we have religion?" many would reply, "Because of its practical benefits."

While the practical benefits of science are important, pragmatics should never obfuscate the underlying goal of science, which is to reveal truth. When the quest for truth is replaced with a concern for the practical, science becomes enslaved to utilitarian concerns that are shaped by political and social exigencies. History suggests that when the political and social exigencies are immoral, science becomes a tool of death and destruction. When the political exigencies are to assert power and maintain a dictatorial status quo, science becomes a tool of repression and falsehoods.

Yet, if the goal of science is to reveal truth, more often than not it will enlighten and enrich mankind. Brigham Young taught, "take the whole truth wherever you find it. It is good; claim it, take it to yourself, and cleave to it, for it will do you good."[15] Put differently, when we quest for truth in science, more often than not, the outcome will be good for humanity.

Scientific history suggests that when we quest for truth, practical benefits will follow. Consider, for example, the work of the nineteenth century Austrian monk Gregor Mendel. Mendel discovered the basic laws of genetics while studying peas in a monastery garden. A neighbor or fellow monk might have protested, "He is studying peas! What practical benefits could possibly come from this research?" Although Mendel's quest for truth regarding inherited traits in garden peas may not have qualified as useful knowledge back then, today we recognize the important contribution his research made to modern genetics. His quest for truth in the heritable traits of garden variety peas led to discoveries in genetics that made possible recent breakthroughs in the Human Genome Project.

GOSPEL TRUTH AND ASSUMPTIONS OF NATURALISM

Science and the gospel share the goal of uncovering truth, but to claim that they are in harmony because of this commonality is an oversimplification. At a deeper ontological level involving assertions about what is real, their relationship is much more complex. As mentioned earlier, during the eighteenth century Enlightenment, the scientific community rejected theistic principles such as the idea that the world was created for a divine purpose and that the creator is continually involved in His creations. The shift to an agnostic and atheistic framework popularized problematic assumptions into the scientific model, assumptions that continue to dominate science today.

Let's take a look at the modern scientific model, called naturalism, and expose a few of the problematic assumptions. Specifically, we will examine materialism, universality, and physical reductionism, and consider the extent to which these assumptions conflict with truths found in the restored gospel of Jesus Christ. This information will help Latter-day Saints avoid being deceived by atheistic falsehoods cloaked in the rubric of scientific fact and authority.

Materialism

Materialism is the belief that everything in the world is composed of physical material in one form or another. Even our thoughts and feelings are no more than biological matter in motion, the product of electrical and chemical processes in our physical bodies. Thus, there is nothing immaterial—no immaterial mind and no immaterial spirit. Ardent materialist and author Willem Drees summed up this viewpoint when he said "the natural [physical] world is the whole of reality that we know of and interact with; no supernatural or spiritual realm distinct from the natural world shows up *within* our natural world, not even the mental life of humans."[16]

How does materialism measure up to gospel teachings? To begin with, we know that our physical bodies are part of the natural world. The book of Abraham states that "the Gods formed man from the dust of the ground" (5:7). Because our physical bodies are composed of physical material, we are subject to many of the same laws of nature that govern the physical world.

Of course, we also possess a spirit. The book of Abraham teaches that after forming man from the dust of the ground, God "took his spirit, and put it into him; and breathed into his nostrils the breath of life" (5:7). Moreover, our spirits are composed of a more fundamental, eternal substance called intelligence. We know that we existed in the beginning as intelligence, that primal substance of humanity that "was not created, neither can be" (93:29). Yet materialism denies the existence of non-corporeal essences such as intelligence and spirits because they are not physically tangible. If something is not tangible or measurable, then according to materialism, it is not real.[17]

Our society is paying a price for materialism's wholesale rejection of nonphysical essences. John A. Widtsoe pointed out, "To limit man's search for truth to the material universe . . . implies that there is no other universe, or that man is incapable of exploring spiritual domains."[18] He continued, "Both alternatives are unacceptable to sound thinking. Man and the eternal universe cannot be confined within the limits of materialism."[19] Thus, science's denial of the spiritual is limiting our ability to more fully understanding the nature of the universe.

Furthermore, by limiting humanity's existence to the physical domain, materialism has contributed to the moral decline of our society. Boyd K. Packer explained that scientific philosophies such as materialism "either by insinuation or by declaration . . . reject God as our creator, as our Father, as our lawgiver."[20] He stated,

> No idea has been more destructive of happiness, no philosophy has produced more sorrow, more heartbreak, more suffering and mischief, no idea has contributed more to the erosion of the family than the idea that we are not the offspring of God, but only advanced animals. There flows from that idea the not too subtle perception that we are compelled to yield to every carnal urge, are subject to physical but not to moral law.[21]

By relegating the universe to matter in motion without divine purpose, materialism sends the message that there is no supreme being and nothing to live for beyond the grave; thus it is okay to "eat, drink, and be merry" as long as it doesn't hurt others. It also sends the message that because there is no God there is no fundamental arbiter of moral truth and thus no such thing as sin (i.e., moral relativism).

Contrary to what many physical materialists may think, the failure of science to observe or measure spiritual substance is not sufficient grounds for denying its existence. Philosopher Grover Maxwell reminded us that because our ability to observe and measure an entity is "a function of our physiological makeup, our current state of knowledge, and the instruments we happen to have available," our inability to observe an entity "has no ontological significance whatever."[22] Put differently, our inability to observe spiritual substance in no way implies that spirits do not exist; it just means that we are presently unable to empirically verify the spiritual given our current limitations. The Doctrine and Covenants explains why science has not been able to empirically verify the spiritual domain. It says that spiritual matter "can only be perceived by purer eyes" (131:7). Acquiring "purer eyes" necessary for perceiving spiritual substance is not something that can be produced in a laboratory or taught in a college classroom; it is a gift from the Lord.

Universality

Universality is the belief that universal laws govern the natural world. Regardless of location in space and time, laws continually exert their influence on the physical world. Examples of such laws include the gravitational pull between matter, the electromagnetic force between charged particles, and the strong nuclear force holding the nucleus of an atom together. By learning more about these laws, it is hoped that we will be better able to predict, control, and understand our natural world.

According to gospel teachings, our physical world is governed by natural laws and the Lord is the author of these laws. Doctrine and Covenants 88:42–43 teaches that "he hath given a law unto all things, by which they move in their times and seasons; And their courses are fixed, even the courses of the heavens and the earth, which comprehend the earth and all the planets." The laws governing our natural world are carried throughout the universe by the Light of Christ, "which is in all things, which giveth life to all things, [and] which is the law by which all things are governed" (D&C 88:12–13). This means that the Spirit of the Lord is the source of the attraction between matter (gravity), the force between charged particles (electromagnetism), and the force holding atomic nuclei together (strong nuclear force). If the Light of Christ stopped filling the immensity of space, the laws of nature would cease to exert their influence, life would come to an end, and our world would fall into chaos.

What is uncertain is whether the laws of nature that are carried throughout the universe are truly universal in the sense of being the same for every location in space. Doctrine and Covenants section 88 states that "every kingdom is given a law . . . [with] certain bounds and also conditions" (v. 38), and that there are different laws for different kingdoms, laws celestial, terrestrial, and telestial (vv. 22–29). These verses suggest that the laws of nature are not universal in the sense of being the same for every kingdom.

Evidence that there are different natural laws for different kingdoms is found in scriptures which indicate that celestial bodies are different in kind from terrestrial and telestial bodies. We read in D&C section 88 that in its celestialized state, the earth will be "sanctified . . . and abide the power by which it is quickened" (v. 26), and that celestial beings have bodies that have been "quickened by a portion of celestial glory" (v. 29). Also, descriptive accounts of visitations by heavenly messengers indicate that celestial beings are not limited by the laws of physics that govern our mortal realm. It thus appears that the laws of nature are context specific and unique to the kingdoms in which they apply. There are different natural laws for different kingdoms, just as there are different spiritual laws for different kingdoms.

More important is the issue of determinism. Determinism is the assumption that material entities are governed by universal laws. With regard to inanimate physical matter, this appears to be the case. As far as we can tell, inanimate matter that has been imbued with intelligence never contradicts the laws of nature that have been instituted by the Lord. From accounts of the Creation we find no evidence to suggest that primordial matter disobeyed the Gods. Abraham 4:18 tells us that during the creation of the world, "the Gods watched those things which they had ordered until they obeyed." The elements obeyed and continue to obey the natural laws that the Lord has established for our world.

Concerns arise when the notion of determinism is applied to human beings. Determinism assumes that because our bodies are composed of physical material that is similar to inanimate physical matter, we are wholly determined by the laws of nature. From this standpoint, any concept or experience of free will is simply an illusion. Cornell University biologist Will Provine expressed this viewpoint:

> Humans are comprised only of heredity and environment, both of which are deterministic. . . . There is simply no room for the traditional concepts of human free will. That is, humans do make decisions and they go through decision-making processes, but all of these are deterministic. So from my perspective as a naturalist, there's not even a possibility that human beings have free will.[23]

The philosopher Bertrand Russell similarly wrote,

> Materialists used the laws of physics to show, or attempt to show, that the movements of human bodies are mechanically determined, and that consequently everything that we say and every change of position that we effect fall outside the sphere of any possible free will. If this be so, whatever may be left for our unfettered volitions is of little value.[24]

The problem with these viewpoints is that they contradict the main purpose of the plan of salvation, which is to make us responsible for our actions so that we can be judged by God and proven worthy of exaltation. We can only be held accountable for our actions (good or bad) if we have moral agency, which is the power to choose between good and evil. Moral agency is not an emergent quality resulting from millions of years of human evolution as some have supposed; it is a divine gift. We were endowed with the gift of agency in the premortal existence and that gift continues with us in mortal life (D&C 29:36; Alma 13:3). Moral agency coupled with human reason gives us a power of choice that is not possessed by any other living creature.[25] Contrary to the assumption of determinism, which underlies universality, we have the ability to act for ourselves and choose the spirit we wish to obey, whether it be good or evil, according to the great plan of God (2 Nephi 2:26).

Physical Reductionism

Physical reductionism assumes that the secrets of nature and life are found in its smallest parts. Thus to truly understand the world, we must gain a fuller understanding of things like molecules, atoms, electrons, and quarks. Certainly reductionism has proven very beneficial in unlocking mysteries of the physical world, but concerns arise when reductionism is applied in an unrestrained manner towards human nature and experience.

When physical reductionism is applied to humanity, as is so often the case, every characteristic we possess is assumed to derive from our biology. Thoughts, feelings, and emotions do not exist in and of themselves; rather they are emergent properties of biological processes. Human Genome Project Nobel Laureate Francis Crick summed up this approach when he claimed, "Your joys and your sorrows, your memories and your ambitions, your sense of personal identity and free will, are in fact no more than the behavior of a vast assembly of nerve cells and their associated molecules."[26]

Notwithstanding the obvious role that biology plays in human conscious experience, *the seed of human consciousness is the spirit, not the body.* The scriptures reveal that we were sentient beings before we were born. Many of us "received [our] first lessons in the world of spirits" which prepared us "to come forth in the due time of the Lord to labor in his vineyard for the salvation of the souls of men" (D&C 138:56). Also, many of us were also called to a holy calling in the premortal existence "on account of . . . exceeding faith and good works" (Alma 13:3). These first lessons and callings, and that fact that they accompany us into mortality, attest that consciousness originates in the world of spirits.

When our spirits entered our physical bodies, consciousness became a product of the interaction between the spiritual and physical. As long as we are in mortality, physical experiences have the power to influence our consciousness. Following death, consciousness will accompany our spirits back to the world of spirits until our spirits are reunited to our physical bodies in the resurrection (Alma 34:34). Clearly, the secrets of human consciousness are not wholly contained within the elements of the physical body as physical reductionism supposes.

Pioneering research on the localization of brain functions by the eminent neurosurgeon Wilder Penfield provides scientific evidence for the existence of a spirit. Penfield was attempting to identify the origins of epileptic seizures by stimulating exposed regions of patients' brains with an electrode. If the initial seizure location could be identified then he would consider removing tissue at the trigger site. By repeatedly stimulating brain regions in conscious patients and noting the effects, Penfield was able to construct a remarkably detailed map of localized functions in the brain. Equally impressive was what he did not find. In all his work on stimulating the human brain, Penfield could not locate the mind.

When Penfield carried out his investigations, patients would report all sorts of sensations, memories, and movements, but the electrode never activated the patients' mind. He could not stimulate the brain and cause a patient to make a choice, to believe something, or to reason. This discovery led him to conclude that "it will always be quite impossible to explain the mind on the basis of neuronal activity in the brain."[27]

Penfield noted that throughout his research, the mind was manifested in the patients' reports of what his electrode caused them to do and feel. For example, when the electrode caused a hand to move, the patients did not say, "I wanted to move my hand"; they said, "I didn't do that, you did."²⁸ He concluded, "The patient's mind, which is considering the situation in such an aloof and critical manner, can only be something quite apart from the neuronal reflex action."²⁹

When Penfield began his studies of the human brain, he had hoped to discover how the brain affects the mind. However, unable to find the physical correlates of the mind in the brain and at the same time ever aware of the presence of mind during his research, Penfield reached an unexpected conclusion. He determined that a human spirit must be the source of the mind. This conclusion brought him great joy. "What a thrill it [was]," he declared, "to discover that the scientist, too, can legitimately believe in the existence of the spirit!"³⁰

CONCLUSION

There are both agreements and disagreements between gospel teachings and naturalism's assumptions. The agreements tend to converge at an understanding of the physical world, while the disagreements largely center on an understanding of humanity. Assumptions that oppose gospel teachings on free will and spirituality need to be understood as merely conjectures, or best guesses, of the way the world operates. If science welcomes open debate over these assumptions, then an awareness of their limited applicability to human existence will surely increase. This increased awareness would likely lead to an improved understanding of what is real and a greater appreciation for alternative, non-scientific ways of knowing.

Notes

1. Karl Popper, *The Logic of Scientific Discovery* (New York: Routledge, 2004), 90.
2. Brigham Young, *Discourses of Brigham Young*. Selected and arranged by John A. Widtsoe (Salt Lake City: Deseret Book, 1954), 38.
3. Ibid., 33; emphasis added.
4. Richard Westfall, *The Life of Isaac Newton* (Cambridge: Cambridge University Press, 1994), 40.
5. Amir D. Aczel, *God's Equation: Einstein, Relativity, and the Expanding Universe* (New York: MJF Books, 1999), 28.
6. Richard Westfall, *The Life of Isaac Newton* (Cambridge: Cambridge University Press, 1994), 60.
7. Amir D. Aczel, *God's Equation: Einstein, Relativity, and the Expanding Universe* (New York: MJF Books, 1999), 28.
8. George F. Richards, Conference Report, October 1933, 112.
9. *Discoveries and Opinions of Galileo.* Translated with an introduction and notes by Stillman Drake (New York: Doubleday Inc., 1957), 59.

10. Ibid., 149.
11. Ibid., 143.
12. Ibid., 143.
13. Ibid., 140.
14. Ibid., 144.
15. Brigham Young, *Discourses of Brigham Young*, selected and arranged by John A. Widtsoe (Salt Lake City: Deseret Book Co., 1954), 152.
16. William Drees, *Religion, Science and Naturalism* (Cambridge: Cambridge University Press, 1996), 12; emphasis added.
17. Brian Greene, *The Fabric of the Cosmos* (New York: Vintage, 2004), 99.
18. John A. Widtsoe, *Evidences and Reconciliations* (Salt Lake City: Improvement Era, 1960), 14.
19. Ibid.
20. Boyd K. Packer, *Things of the Soul* (Salt Lake City: Bookcraft, 1996), 111.
21. Ibid.
22. Grover Maxwell, *The Ontological Status of Theoretical Entities, In Philosophy of Science: The Central Issues*. Edited by M. Curd and J. A. Cover (New York: W. W. Norton and Company, 1998), 1062.
23. In Russell Stannard, *Science and Wonders: Conversations about Science and Belief* (Boston, MA: Faber and Faber, 1996), 60.
24. Bertrand Russell, *Russell on Religion: Selections from the Writings of Bertrand Russell*. Edited by L. Greenspan and S. Anderson (New York: Routledge, 1999), 178.
25. See president David O. Mckay, Conference Report, October 1969, 6–7; Boyd K. Packer, *Things of the Soul* (Salt Lake City: Bookcraft, 1996), 110.
26. Phillip E. Johnson, *The Wedge of Truth: Splitting the Foundations of Naturalism* (Downers Grove, IL: InterVarsity Press, 2000), 122.
27. Wilder Penfield, *The Mystery of the Mind* (Princeton, NJ: Princeton University Press, 1975), 80.
28. Ibid., 76.
29. Ibid., 55.
30. Ibid., 85.

NINE
BRINGING GOSPEL TRUTH TO SCIENCE

We shall overcome any seeming contradictions between science and religion if and as we learn to adhere to the teachings of the restored Church, which have always been dedicated to the search for knowledge and intelligence, which is the Glory of God.

—HUGH B. BROWN

In 2000 I attended the American Psychological Society convention in Portland, Oregon. I listened to a keynote address by a renowned neuroscientist who said that he was confident that human experiences would one day be fully explained by nervous system activity. His comment implied that all human thoughts, feelings, and emotions were reducible to biological events. While I respected his accomplishments, it was apparent that we had vastly different assumptions about the fundamental nature of humankind.

Many Latter-day Saints face similar contradictions between scientific and religious beliefs. We hear one thing from a religious point of view, and hear something different from a scientific point of view. Is a reconciliation between teachings in the restored gospel and science possible, and if so, what and who must change?

RECONCILIATION

Achieving a greater harmony between science and our religion requires that the scientific community change its assumptions to recognize basic gospel truths. This process involves (a) acknowledging that

God is the source of natural laws and supernatural miracles; (b) recognizing that the spirit is the life force and source of human consciousness, and that consciousness is not reducible to physical matter; (c) acknowledging that humanity is endowed with moral agency, which gives us the power to choose between good and evil; and (d) recognizing that our world and the universe are the handiwork of Him who created and sustains everything.

It is interesting to note that these beliefs were popular during the Scientific Revolution of the sixteenth and seventeenth centuries, thanks to men like Galileo Galilei, Isaac Newton, and Robert Boyle. In a way, then, we need to return to the scientific-religious relationship that existed during the Scientific Revolution. To bring about this change, we must be willing to affirm our basic convictions about God, agency, and spirituality in classrooms, papers, scholarly presentations, and scientific discussions, when appropriate. If this makes us feel uncomfortable, it is only because we have been led to believe that religious beliefs have no place in science.

The idea that religious beliefs should be kept out of science is part of a grand deception that began during the Enlightenment and is now being perpetuated by secular humanists. Secularists would have us believe that to be scientific one must embrace a separation of science and religion; this is simply not true. Some of the greatest scientific minds of all time, such as Newton, Galileo, Descartes, and Boyle, repeatedly mentioned basic Christian beliefs in their writings, demonstrating that it is possible to refer to such beliefs without diminishing scientific rigor. Even the late Harvard zoology professor and author Stephen J. Gould mentioned God and scripture so often in his writings that he sounded more like a deist than the agnostic he claimed to be.[1] If gospel truths correspond with issues that science is addressing, then we should be willing to bring that truth to the world in an unpretentious manner.

Mentioning God, agency, and the human spirit in scholarly settings will bring derision from some; nevertheless, we must stand up for these beliefs if we hope to create a greater harmony between the gospel and science. I believe that this is a reason Brigham Young counseled us to "be a people of profound learning pertaining to the things of the world."[2] In order to promote a greater harmony between the gospel and science, we need to acquire both secular and spiritual knowledge. Knowledge from both areas will enable us to reconcile differences and reinforce similarities.

Consider, for instance, the debate about whether the earth was created out of nothing (ex nihilo) or from preexisting materials. Many theologians believe in creation ex nihilo, while many scientists argue that it is impossible to create something out of nothing. On this question, science is in agreement with gospel truth. Latter-day scripture points out that this world was organized from pre-existing material. During the Creation, the Gods declared, "We will go down, for there is space there, and we will take of these materials, and we will make an earth whereon these may dwell" (Abraham 3:24). And in the Doctrine and Covenants 93:33 we read that "the elements are eternal."

We can achieve a greater harmony between the gospel and science by emphasizing such similarities. Brigham Young did just that when he said,

> You may take geology, for instance, and it is true science; not that I would say for a moment that all the conclusions and deductions of its professors are true, but its leading principles are; they are facts—they are eternal; and to assert that the Lord made this earth out of nothing is preposterous and impossible. God never made something out of nothing; it is not in the economy or law by which the worlds were, are, or will exist.[3]

Another example includes the commonly held belief, among other religions, that the earth was created in six 24-hour days. This time period conflicts with the geological record which indicates that it took much longer for the earth to evolve to its current state. Once again, gospel teachings are consistent with geology's findings. Abraham's use of the word *time* in his account of the Creation is commonly believed to refer to creative periods rather than twenty-four-hour days. Bruce R. McConkie stated that a "day" in the Creation is "a specified time period; it is an age, an eon, a division of eternity; it is the time between two identifiable events . . . the duration needed for its purposes."[4]

Brigham Young said, "Our religion will not clash with or contradict the facts of science in any particular."[5] This puts us Latter-day Saints in the unique position of being able to reconcile science and Christianity. When science has its ideas correct, those ideas will agree with similar teachings in our gospel. By voicing the similarities between LDS and scientific beliefs, we are promoting a more harmonious relationship between the gospel and science.

SOLVING THEOLOGICAL AND PHILOSOPHICAL CONCERNS

There are other ways in which truths in the restored gospel can benefit secular scholarship. For centuries scholars have been perplexed by the mind-body problem, the concept of divine intervention, and the problem of evil. Mankind's inability to satisfactorily resolve these theoretical and theological concerns has widened the divide between science and Christianity. Yet confusion surrounding these issues largely stems from an incomplete understanding of the nature of the human soul, divine laws, and the plan of salvation. As I will show, truths found in the gospel of Jesus Christ can help resolve these concerns, illustrating once again the potential of the gospel to bring about a greater harmony between Christianity and science.

Mind–Body Problem

The mind-body problem refers to the belief in an ontological dualism, that there is a physical and nonphysical reality. This dualism has created a problem for those trying to explain how an immaterial mind or spirit can interact with a material physical body. Efforts to adequately explain how a physical entity interacts with a nonphysical entity have largely proven unsatisfactory, as evidenced by the following attempts which leave many unanswered questions.

Psychophysical parallelism offers a solution to the mind/spirit-body problem by proposing that physical and nonphysical entities coexist independent of each another. A major drawback with this proposal is that it does not agree with the commonly held belief that the spirit and physical body interact to form a unified soul, nor does it explain how the spirit is able to influence the body, and vice versa. Another proposed solution, epiphenomenalism, claims that the nonphysical mind is simply a manifestation of biological processes in the brain. While this notion may satisfy a good many atheists, it denies our experience of thoughts influencing our bodies, and it reduces meaningful psychological processes to nothing more than neurological activity. Emergent interactionism, on the other hand, attempts to solve the mind/spirit-body problem by claiming that psychological experiences emerge from the brain, and that once they have emerged, they are real in the sense of that they can influence the physical body. This account is unsatisfactory because it leaves unanswered the question of how something nonphysical might influence and emerge from something physical.

Unable to provide a tenable explanation for dualism, some dualists have resigned to the notion that because God is able to do all things, He is able to accomplish the seemingly impossible and unite the spirit and body into a unified whole. While it is true that "the Lord is able to do all things according to his will, for the children of men" (1 Nephi 7:12), it is also true that the Lord has given us some indication of how the spiritual and physical interact.

The Doctrine and Covenants states, "There is no such thing as immaterial matter. All spirit is matter, but it is more fine or pure" (131:7). Hence, the solution to the mind/spirit-body problem lies in recognizing that the sentient, nonphysical spirit is composed of a substantive material that occupies space, and as such is capable of interacting with a material physical body. There is no need to speculate on how something immaterial can interact with something material because both the spirit and physical body are material. The concept of a material spirit and material physical body uniting to form a unified soul provides a legitimate solution to the mind-body problem that has perplexed scholars for ages.

Divine Intervention

Supernatural intervention refers to divine causation that operates outside the bounds of natural laws governing our world. Critics argue that accepting supernatural intervention would undermine the integrity of science because to accept supernatural events would mean having to accept that natural laws are violable or that they may be temporarily interrupted. Such concepts are incompatible with naturalism's assumption of a law-governed world.[6]

Of course, naturalism's denial of supernatural intervention suits the agnostics and atheists just fine. The deists also have no difficulty accepting this position because it is consistent with their belief in an uninvolved creator. However, for theists who accept revelation, miracles, and redemption, rejecting supernatural intervention because it violates the laws of nature is unacceptable. But how can we reconcile supernatural intervention in a world governed by natural laws?

One proposal involves assimilating scientific explanations "within the framework of an all-encompassing, biblically informed, theistic world view."[7] Theists would incorporate atheistic scientific assumptions into their theistic world view with the hope that the two belief systems would eventually merge in a logical fashion. In my opinion, this approach is akin to introducing a wolf into a flock of sheep with the hope that they will get along. This sort of thing was tried a few centuries ago. Eighteenth century Enlightenment values were incorporated into the seventeenth century theistic world view, resulting in the secularization of science.

Another attempt at reconciling natural and supernatural processes comes from process theology. According to process theologians such as Alfred Whitehead North and David Ray Griffin, divine intervention is manifested in the everyday natural processes that sustain the world, as suggested by the scripture which claims that God is in all things and all things are in God.[8] In other words, "miracles" are manifested through the divine power that maintains everything.

While it is true that the power of the Spirit of the Lord to maintain life throughout the universe is a miracle, it is also true that limiting miracles to everyday natural processes undervalues the significance of supernatural miracles. To process theologians, biblical accounts of supernatural miracles are just metaphorical descriptions of everyday, natural processes. For example, the biblical account of the creation of Adam and Eve is considered a fanciful account of the origin of life. They replace the Creation story with a divinely implemented process of human genesis consistent with evolutionary principles.

The key to reconciling supernatural and natural processes lies in recognizing that divine intervention does not violate natural laws. We learn in the scriptures that through adherence to principles of righteousness, glorified beings have power over natural elements and processes (D&C 130:20–21). This power is evident in the Savior's declaration: "I have created by the word of my power, which is the power of my spirit" (D&C 29:30).

Moreover, we know that the celestial kingdom is not governed by the same laws that govern our telestial kingdom (D&C 88:38). Hence, miracles do not break, interrupt, or violate our natural laws because the Lord is not beholden to the same boundaries and conditions that govern our world. When miracles occur, He is exercising His power according to His celestial laws on behalf of His children in a lower kingdom.

When something amazing occurs that cannot be explained by science, believers are inclined to say that God intervened and nonbelievers are inclined to say that it was a chance occurrence. Alvin Plantinga asks, "Where is it written that [a supernatural] conclusion can't be part of science?" and "Why couldn't one conclude this precisely as a scientist?"[9] He provides the answer by saying that there is absolutely nothing in science to preclude one from reaching the conclusion that miracles are real. I wholeheartedly agree. There are no rational grounds for denying miracles in a law-governed world. Denial of miracles in the scientific community results from unbelief rather than concern for the violation of natural laws.

The Problem of Evil

The problem of evil is the dilemma of how a perfectly good, all knowing, and all powerful creator could allow evil and suffering in the world. If a supreme creator were omnipotent and omniscient, skeptics claim, then surely he would intervene to eliminate pain and suffering. Because there is so much pain and suffering, they argue, there must not be a supreme being. According to the process theologian David Ray Griffin,

> As long as the idea of a Divine Creator is identified with the God of supernaturalistic theism, who could but does not prevent evils such as birth defects, cancer, and the Holocaust, the arguments for God, such as the arguments from order and beauty, are canceled out by the argument against God posed by the world's disorder and evil.[10]

Those who deny God's existence and miracles because of evil and suffering do not understand the role of agency in the plan of salvation. Agency is a fundamental principle in Heavenly Father's plan to bring to pass the eternal life of man. Agency allows us to prove ourselves worthy of exaltation. After choosing the materials wherewith to make the earth, the gods declared, "And we will prove them herewith, to see if they will do all the things whatsoever the Lord their God shall command them" (Abraham 3:25).

But agency alone is not a sufficient condition for proving ourselves worthy of exaltation; proving ourselves worthy of exaltation also requires the existence of evil. In the premortal existence Lucifer and his followers exercised their agency to rebel against Heavenly Father and have since become a source of temptation and evil on earth. Being enticed by the adversary to do evil, and being enticed by the Spirit of the Lord to do good, is essential to the plan of salvation. The Book of Mormon prophet Lehi pointed out that "man could not act for himself save it should be that he was enticed by the one or the other" (2 Nephi 2:16). Without evil there would be no choice, thus making it impossible for us to choose righteousness and qualify for the blessings of exaltation.

Lehi also taught that without evil and suffering there could be no righteousness and happiness. Without righteousness and happiness there could be no God, and "if there is no God, we are not, neither the earth; for there could have been no creation of things, neither to act nor to be acted upon; wherefore all things must have vanished away" (2 Nephi 2:11). And so agency and evil are essential conditions for the existence of the earth and the universe in which we live. This does not mean that God is the source of evil and suffering. The scriptures testify that "all things which are good cometh of God; and that which is evil cometh of the devil" (Moroni 7:12). Evil and suffering result when people exercise their moral agency to follow the enticings of the evil one.

Finally, we do not know all the reasons people suffer from circumstances that are beyond their control. We know that we will be tried and tested and that how we exercise our agency while undergoing trials plays an important part in qualifying for exaltation. The Lord said, "My people must be tried in all things, that they may be prepared to receive the glory that I have for them, even the glory of Zion; and he that will not bear chastisement is not worthy of my kingdom" (D&C 136:31). Trials and suffering may therefore be thought of as tests to determine whether we will continue in righteousness and trust in the Lord, regardless of our circumstances. Alma testified that if we endure trials in faith, the Lord will support us in our afflictions and we will be "lifted up at the last day" (36:3).

CONCLUSION

A complete harmony between the gospel of Jesus Christ and science will likely come during the Millennium. I believe that during the Millennium, science will be harmonized with the gospel in much the same way that governments will be harmonized into the political kingdom of God. Brigham Young taught that during the Millennium, the political kingdom of God "grows out of The Church of Jesus Christ of Latter-day Saints, but it is not the church."[11] In much the same way, science will grow out of the Church, but will not be the Church. Until then, when all truth is circumscribed into one whole, we can rely on gospel teachings and inspiration from the Spirit of the Lord to expand our understanding of secular truth.

Notes
1. See for example, Stephen J. Gould, *The Living Stones of Marakech: Penultimate Reflections in Natural History* (New York: Harmony Books, 2000).
2. Brigham Young, *Discourses of Brigham Young*. Selected and arranged by John A. Widtsoe (Salt Lake City: Deseret Book Co., 1954), 254.
3. Ibid., 258.
4. Bruce R. McConkie, *Sermons and Writings of Bruce R. McConkie* (Salt Lake City: Bookcraft, 1998), 181.
5. Brigham Young, *Discourses of Brigham Young*. Selected and arranged by John A. Widtsoe (Salt Lake City: Deseret Book, 1954), 258.
6. William Drees, *Religion, Science and Naturalism* (Cambridge University Press, 1996), 94–95.
7. David Ray Griffin, *Religion and Scientific Naturalism: Overcoming the Conflicts* (State University of New York Press, 2000), 56.
8. Ibid., 97.
9. Ibid., 49.
10. Ibid., 54.
11. Hoyt W. Brewster, Jr., *Behold, I Come Quickly: The Last Days and Beyond* (Salt Lake City: Deseret Book Co., 1994), 211.

TEN

GOSPEL TRUTH AND QUANTUM MECHANICS

There is the whole field of subatomic physics; the unpredictable and uncontrolled nature of such things obliges us to classify them as the unexplained or the miraculous. Of course, for us the important breakthroughs are the dispensations of the gospel, which can come only by the opening of the heavens.

—Hugh Nibley

Discoveries in quantum mechanics during the early twentieth century introduced provocative ideas that challenge traditional concepts about the nature of the universe. Despite its strangeness, quantum mechanics makes clear assumptions about the world. We'll examine three ontological assumptions in quantum mechanics (non-locality, randomness, and wave-particle duality) and consider the extent to which these assumptions agree or disagree with gospel teachings. Regardless of one's level of scientific knowledge, most readers will be able to follow this thought provoking discussion with relative ease. I hope that this discussion will generate interest in the relationship between the gospel and theories of science within the LDS community.

QUANTUM NON-LOCALITY

People generally believe that events or forces must travel through space and time in order to influence another event in a different location. Imagine that you have a friend who is an archery expert. She is going to demonstrate her archery skills by popping a balloon thirty-six meters away. After you position the balloon on a target and move back thirty-six meters, she sets the arrow, aims, and fires. For a split second you see the arrow travel through space and then pop the balloon as it strikes the target. This linear sequence of events is called locality. An arrow is fired, it travels through space and time to the location where the target is positioned, and pops the balloon. Locality is so common to human experience that we generally expect all causal events to follow a linear sequence.

However, recent discoveries in quantum mechanics challenge the concept of locality. Research has shown that certain events do not have to traverse space and time to influence other events. Events may be interconnected in such a way that, regardless of the distance separating them in space, they simultaneously occur. Studies of the spin characteristics of photons demonstrate this interconnectedness. When measured on a certain axis of spin, photons move in either a clockwise or counterclockwise direction. In the early 1980s, Alain Aspect discovered that when twin photons emitted from a calcium atom are sent traveling in opposite directions (up to thirteen meters apart), measurements taken of one photon's spin instantaneously influence the spin of its twin.

More recently, in 1997 Nicolas Gisin from the University of Geneva discovered that twin photons emitted eleven kilometers apart, an astronomical distance for their size, demonstrate the same interconnectedness; measurement of one photon's spin instantly influenced the spin of its twin. The interconnectedness of photons' spin characteristics suggests that the photons are part of the same system, regardless of the distance separating them. These findings led physicist and author Brian Greene to conclude that "what you do at one place can be [instantaneously] linked with what happens at another place."[1] Currently many physicists limit this phenomenon to events that do not involve the transmission of information, energy, and matter.

Gospel doctrine and personal experience indicate that not only are instantaneous (non-local) processes possible, but they regularly involve the transmission of information. When we pray, we are engaging in a type of instantaneous, non-local communication with deity. The scriptures point out that when we pray to Heavenly Father, our communication is instantly received. Our words and thoughts immediately traverse space and time to the dwelling place of God. The same is true when Heavenly Father answers our prayers (for example, see 3 Nephi 1:11–13 and Proverbs 15:29). What makes this non-local communication possible is the Light of Christ. The Spirit of the Lord, "which fills the immensity of space," is the means by which God instantly knows our thoughts and feelings and is the medium by which He instantaneously communicates with His children (D&C 88:6–13).

Non-local processes involving the instantaneous transportation of physical material are also a reality. Brigham Young stated that a personage who has achieved a certain level of perfection can "waft himself to the moon or to the North Star, or to any other of the fixed stars, *and be there in an instant*, in the same manner that Jesus did when he ascended to the Father in heaven and returned to the earth again."[2] Non-local interstellar travel makes possible instantaneous visitations by angels, something I imagine Nephi was grateful for when his brothers Laman and Lemuel were beating him outside Jerusalem. As mentioned earlier, glorified beings are not bound by mortal laws governing our telestial space and time. They live by laws unique to the celestial kingdom in which they reside.

RANDOMNESS

Another intriguing discovery in quantum mechanics is the apparent randomness of certain subatomic particles. In the aforementioned example of photon spin, research suggests that the spin of photons is not determined by laws of nature; it is decided randomly. Randomness is also observed when light photons encounter semi-reflective surfaces (like sunglasses) that have been designed to transmit some photons and reflect others. Studies show that the decision to transmit or reflect a single photon of light is not determined by natural laws; it is decided randomly. This evidence led physicist Paul Dirac to conclude that "nature makes a choice."[3]

Interestingly, quantum randomness contradicts scientific naturalism's assumption of determinism, wherein matter is thought to be wholly governed by the laws of nature. Albert Einstein was troubled by quantum randomness because it contradicted his belief in a universe that is strictly determined. His famous statement, "I shall never believe that God plays dice with the world,"[4] reflected his belief that the universe is governed by laws and nothing is left to chance. Thus he concluded that quantum theory failed to give a complete understanding of nature.

John A. Widtsoe held a similar opinion of quantum randomness. He wrote,

> Every process of nature is orderly. Chance, disorder, chaos are ruled out of the physical universe.... The sun does not rise in the east today and in the west tomorrow. That means that the phenomena of nature are products of law. The infinitely large or the infinitely small move in obedience to law. In man's earnest search for truth, no exception to this process has been found. Apparent deviations, such as the famous uncertainty principle operating in the subatomic world are but expressions of man's incomplete knowledge, which always disappear with increasing knowledge.[5]

John A. Widtsoe's skepticism regarding randomness in nature is consistent with scripture which teaches that the Lord "hath given a law unto all things, by which they move in their times and their seasons" (D&C 88:12). These laws are carried throughout the universe by the Light of Christ, which "proceedeth forth from the presence of God to fill the immensity of space" (D&C 88:12). This scripture suggests that all of the Lord's creations are governed by laws and nothing in nature is left to chance. But what about humanity?

Some have proposed that quantum randomness is a vestige of human agency—that it allows for agency in a determined world.[6] This proposed connection is an intriguing hypothesis, but I do not think that quantum randomness is the foundation of human agency. Moral agency and quantum randomness differ significantly. Randomness refers to chance occurrence while moral agency refers to the ability to choose between good and evil. We know that our spirits possessed moral agency in the premortal existence and that this agency gave us the power to choose the Savior's plan of salvation. This premortal agency suggests that the foundation of free will lies with the spirit, and not in subatomic physical particles. Moreover, if quantum randomness were the foundation of moral agency, then we would expect entities such as animals, trees, and rocks to have moral agency because their subatomic particles are capable of quantum randomness. Of course, this is not the case; human beings are the only entities endowed with moral agency.[7]

WAVE-PARTICLE DUALITY

The last quantum phenomenon that we'll consider is observation creating the reality of the thing being measured. This phenomenon is described by the wave-particle duality theory of electrons. To understand this theory, imagine that we've set up an electron gun capable of firing a continuous beam of electrons toward an opaque panel. In the center of the panel there are two very narrow slits positioned side by side. On the other side of the panel is a screen which registers the impact location of electrons that pass through the two slits in the panel.

When we turn on the gun and fire a continuous beam of electrons toward the panel, what kind of pattern will be visible on the screen? Common sense suggests that we would see two bright bands behind the place where the particles passed through the slits. But this is not what happens. Instead there would be a pattern of alternating, vertical dark and bright bands that diminish the greater their distance from the center of the image. This is called a wave interference pattern. It is similar to the wave pattern of alternating peaks and troughs created when a pebble is tossed into a pool of water.

Schrodinger believed that this wave pattern was caused by the combined behavior of individual electrons. However, this explanation was refuted by the following modification to the previous experiment. Instead of firing a continuous beam of electrons toward the panel, we adjust the gun so that it fires one electron every few seconds. If we left the gun firing single electrons every few seconds and came back the next day, we would find an interference wave pattern identical to the pattern we observed when the gun fired electrons continuously.[8] This means that the interference wave pattern is not the result of the combined behavior of electrons, as Schrodinger had supposed. But how can the actions of independent electrons combine to form an interference wave pattern?

A very intriguing answer to the above question was provided by physicist Max Born. Born proposed that the wave pattern produced by single firing electrons was not created by the actual electrons themselves, but by immaterial potentialities representing the probable location of the electrons on the screen. According to Born, after an electron leaves the gun, it no longer exists as a particle. It becomes an immaterial potentiality, a wave representing probable locations where the electron will end up. As the electron's probability wave passes through the slits, it divides into two probability waves that interact or interfere with one another on the other side of the screen, thus creating an interference wave pattern.

Bright peaks and troughs in the interference wave represent locations in space where the electron will most likely reappear, and dark regions (caused by waves canceling each other out) represent locations in space where the electron will not reappear. After the interference pattern has formed, the electron then emerges from its metaphysical domain of wave potentialities and once again becomes a real particle when its position is measured on the screen. Colgate University physicist Shimon Malin described this phenomenon as such:

> When the electron is not measured—in the space between the electron gun and the screen, for example—*it does not exist at all as an actual "thing." It exists merely as a field of potentialities.* Potentialities for what? For becoming an actual "thing," having certain properties, *if measured.* But as long as it is not measured, there is no thing there. We are suggesting that the electron actually exists at the electron gun and that it actually exists at the TV screen, but *it does not exist in between, except as a collection of potentialities.*[9] (emphasis added)

The concept of something not existing except when measured has intriguing implications for our physical world. To begin with, this phenomenon fueled the belief among many physicists that something must be measurable in order to exist. As I pointed out in a previous chapter, this antirealism stance is troubling. It limits acceptance of what is real to that which is perceived by the physical senses (i.e., materialism), thereby restricting our ability to capture the essence of what is real in the universe, including spiritual entities.

The claim that subatomic entities lie in a state of metaphysical abeyance until an observer's attention snaps them into existence also raises the centuries-old debate of whether something must be perceived in order to exist. As a realist who believed in a physical reality independent of human consciousness, Albert Einstein was deeply troubled by such claims. He rejected the notion that being requires perceiving. He quipped, "When no one is observing the moon, is it still there?"[10]

You have addressed this issue if you've ever contemplated this question: If a tree falls in the forest and no one hears it, does it still make a sound? The next time someone asks you this question you might want to reply: "Does the tree even fall or exist if no one ever witnesses it?" Two eighteenth century philosophers who would have answered no to the latter question were the skeptic David Hume and the immaterialist George Berkeley. Hume maintained that belief in the existence of an external world that is independent of the human mind is an illusion, and Berkeley's dictum *esse est percipi* ("to be is to be perceived") claimed that only that which is perceived is real. More recently, process philosophers in the tradition of Alfred North Whitehead have similarly downplayed the existence of an independent, objective world. Process philosophers generally believe that the fundamental constituents of reality are not particles of nature, but rather occasions of human experience. For them, reality is grounded in experience.

Regardless of whether something must be perceived in order to exist, God's omniscience ensures that all things are continuously perceived. The Light of Christ which fills the universe places all things before the awareness of God (D&C 88:41). His Light makes it possible for Him to know at an instant when a sparrow falls out of a tree and how many hairs are on the tops of our heads, notwithstanding having created worlds without number.

Therefore, the notion that subatomic particles slip in and out of existence based on whether they are being measured is likely a manifestation of the limitations of quantum theory. Albert Einstein and John A. Widtsoe agreed. They believed that quantum theory was an incomplete description of the subatomic world. The eminent physicist Richard Feynman declared, "I think I can safely say that nobody understands quantum mechanics."[11] Indeed, it seems we have much more to learn about the quantum world.

Notes

1. Brian Greene, *The Fabric of the Cosmos* (New York: Vintage, 2004), 114–115.
2. Brigham Young, *Discourses of Brigham Young*. Selected and arranged by John A. Widtsoe (Salt Lake City: Deseret Book Co., 1954), 424; emphasis added.
3. Shimon Malin, *Nature Loves to Hide: Quantum Physics and the Nature of Reality, a Western Perspective* (New York: Oxford University Press, 2001), 132.
4. Amir D. Aczel, *God's Equation: Einstein, Relativity, and the Expanding Universe* (New York: MJF Books, 1999), 120.
5. John A. Widtsoe, *Evidences and Reconciliations* (Salt Lake City: Improvement Era), 19.
6. See Philipp G. Frank, "The Variety of Reasons for the Acceptance of Scientific Theories," in *Introductory Readings in the Philosophy of Science*. Edited by E.D. Klemke, R. Hollinger, and A. D. Kline (New York: Prometheus Books, 1988).
7. See president David O. Mckay, Conference Report, October 1969, 6–7; Boyd K. Packer, *Things of the Soul* (Salt Lake City: Bookcraft, 1996), 110.
8. Timed photos of such an experiment are available in the 1976 American Journal of Physics, 66, 306.
9. Shimon Malin, Nature Loves to Hide: *Quantum Physics and the Nature of Reality, a Western Perspective* (New York: Oxford University Press, 2001), 48.
10. Brian Greene, *The Fabric of the Cosmos* (New York: Vintage, 2004), 103.
11. Brian L. Silver, *The Ascent of Science* (New York: Oxford University Press, 1998), 357.

ELEVEN
COMING TO THE KNOWLEDGE OF THE TRUTH

A firm, unchangeable course of righteousness through life is what secures to a person true intelligence.

—Brigham Young

"NEVER COMING TO THE KNOWLEDGE OF THE TRUTH"

The prophet Daniel prophesied that "knowledge shall be increased" in the last days (Daniel 12:4). We see evidence of this in the restoration of the fulness of the gospel and in the increase of secular knowledge since the second illumination. We've seen that these blessings are a direct result of the Spirit of the Lord being poured out upon the earth in the last days. Yet if we do not receive this blessing of knowledge in the way that the Lord intended it to be received, it may be turned to our disadvantage. The Apostle Paul knew that the blessing of increased knowledge would become a curse for many not willing to follow the commandments of the Lord. He prophesied that in our time, people would be "ever learning, and never able to come to the knowledge of the truth" (2 Timothy 3:7).

Today's electronic technologies have placed an abundance of information at our disposal. There is so much information available that we could spend our whole lives collecting new facts and ideas. However, as the Apostle Paul warned, the pursuit of knowledge may become a vice if it diverts our attention from the weightier matters of life and salvation. We may become so preoccupied with acquiring secular knowledge that we fail to obtain the most important knowledge of all—an understanding of the doctrines of salvation in the gospel of Jesus Christ. The scriptures remind us that "to be learned is good, if [we] hearken unto the counsels of God" (2 Nephi 9:29). If we are well-educated but do not study the counsels of the Lord, we are at risk for becoming prideful and distancing ourselves from Heavenly Father.

Many scholars who lack an understanding and conviction of gospel principles worship at the altar of the god of science. With the book of philosophy and science as their bible, they think that they have obtained lasting happiness and wisdom in their scholarly work. They are mistaken. The scriptures testify, "O the vainness, and the foolishness of men! When they are learned they think they are wise, and they hearken not unto the counsel of God, for they set it aside, supposing they know of themselves, wherefore, their wisdom is foolishness and it profiteth them not" (2 Nephi 9:28).

President Thomas S. Monson cautioned us on the costs of unmitigated secular learning. He said,

> The world moves at an increasingly rapid pace. Scientific achievements are fantastic, advances in medicine are phenomenal, and the probings of the inner secrets of earth and the outer limits of space leave one amazed and in awe. In our science-oriented age we conquer space, but we cannot control self; hence, we forfeit peace.[1]

History suggests that those who are ever learning without embracing principles of righteousness are prone to forget their moral obligations to God and humanity. The effects of forgetting one's moral obligations were especially evident during World War II. A handful of Nazi physicians, some of whom held prominent posts in the medical world, committed atrocities in the name of science.

In one study carried out in the Luftwaffe Air Force Medical Corps, prisoners were subjected to decompression and freezing experiments to simulate the effects of air pressure loss at 100,000 feet. The only military personnel who refused to carry out the experiments were Christian medics. Upon hearing their objections, SS Chief Heinrich Himmler wrote a letter to the Luftwaffe field marshal Ernhard Milch, protesting the difficulties caused by "Christian medical circles" and called for a "non-Christian physician" to head up the project.[2]

The effects of "ever learning without coming to the knowledge of the truth" are now manifested in a different, yet equally pernicious, way. The effects are spiritual. Efforts to acquire secular knowledge without understanding principles of righteousness have contributed to the moral decline of society. In a recent example, special interest groups succeeded in getting the American and Canadian Psychological Associations to adopt resolutions assailing traditional family values and condoning sexual deviancy, all in the name of science.

In a speech at the 1948 Armistice Day celebration in Boston, General Omar Bradley commented on the struggle between morality and amorality. He warned:

> Humanity is in danger of being trapped in this world by its moral adolescence. Our knowledge of science has clearly outstripped our capacity to control it. We have too many men of science; too few men of God. We have grasped the mystery of the atom and rejected the Sermon on the Mount. Man is stumbling blindly through a spiritual darkness while toying with the precarious secrets of life and death. The world has achieved brilliance without wisdom, power without conscience. Ours is a world of nuclear giants and ethical infants.[3]

As Latter-day Saints, we need to take a stand in the defense of truth and righteousness, no matter how unpopular it might be. When we uphold principles of righteousness, we bring wisdom to secular scholarship in the form of good judgment, compassion, and sound decision making. John A. Widtsoe highlighted the need for righteousness to guide scholarly endeavors by declaring:

> As science advances and increases, as new discoveries are made, as more complete command is obtained over the forces of nature, the more necessary it becomes that we have a religion to guide us in employing these discoveries. To save the world from science, and to make science the builder of a good world, we must hasten our progress towards the fuller acceptance of God.[4]

QUALIFYING FOR THE ENLIGHTENING POWER OF THE SPIRIT OF THE LORD

The main theme of this book is that the Spirit of the Lord enlightens people's minds and provides them with secular knowledge. The enlightening power of the Spirit operates on certain principles. By learning the principles governing its influence, we may more fully receive its guidance and enjoy what Truman Madsen described as the "ultimate teaching process and the source of genuine learning and intelligence."[5] Let's take a look at a few of these principles.

To receive abundant blessings of knowledge from the Spirit of the Lord we must learn the principles of righteousness. But understanding the principles of righteousness alone is not enough. We must endeavor to live the principles of righteousness. Robert J. Mathews pointed out that the Pharisees and Sadducees had many of the facts of the gospel that the Lord gave to Moses, but they were unable to comprehend the things of God, even when they read the scriptures, because they did not have the Spirit.[6] The Pharisees and Sadducees did not have the Spirit because they did not hearken unto the commandments of the Lord. When we understand and live the principles of righteousness, we qualify ourselves to receive its guidance. Joseph Fielding Smith described the relationship between obedience and enlightenment in the following way:

> The Lord has held in reserve for those who obey him this great blessing of the fulness of truth, and through obedience it is received and in no other way. So with all their searching, with all their delving into the earth and examining of the heavens, man cannot discover the fulness of truth without submission to the principles of the gospel and placing their lives in harmony with the Holy Spirit and walking in obedience to the commandments of the Lord. There is no other way in which all truth may be obtained.[7]

To freely receive the enlightening power of the Spirit we must also "seek out of the best books words of wisdom; seek learning, even by study and also by faith" (D&C 88:118). This scripture suggests that we should search out literature on the issue we're investigating. Many scholarly works contain inspired knowledge that has been preserved for our benefit and the benefit of future generations. Brigham Young taught:

> The revelations of the Lord Jesus Christ to the human family are all the learning we can ever possess. Much of this knowledge is obtained from books which have been written by men who have contemplated deeply on various subjects, and the revelation[s] of Jesus have opened their minds.[8]

Mental discipline and diligent study are also important for receiving enlightenment from the Spirit of the Lord. Brigham Young taught that the greater our mastery over thoughts and feelings, "the faster [we] can grow and increase in the knowledge of the truth."[9] Mental discipline manifests our sincere desire to receive knowledge and "tunes" our minds to the promptings of the Spirit of the Lord. Study and learning are important because they put us in a position to comprehend the new light and knowledge that the Lord is waiting to give us. He gives "unto the children of men line upon line, precept upon precept, here a little and there a little" (2 Nephi 28:30), suggesting that blessings of knowledge come gradually and are built upon preexisting knowledge. The Lord will not give us knowledge that we are not prepared to receive because He will not make us "run faster . . . than [we] have strength" (D&C 10:4).

Humility and faith are also important. The Lord has said, "Let him that is ignorant learn wisdom by humbling himself and calling upon the Lord his God, that his eyes may be opened that he may see, and his ears opened that he may hear. For the spirit is sent forth into the world to enlighten the humble and contrite" (D&C 136:32–33). Those who do not pray because of a lack of faith may still qualify for enlightenment if they reverence the Lord's creations. John A. Widtsoe taught that "there are some truthseekers . . . who do not speak to the Lord directly, but they . . . stand reverently before the power in all things, which is their conception of God."[10] He added, "Prayer as commonly understood, *or its equivalent*, is a requisite for those who are to travel the way to truth."[11]

Finally, we should give thanks to Heavenly Father when the Spirit of the Lord quickens our understanding; but that is not all. Because we have been instructed to return thanks for *every blessing* we receive (Alma 7:23), we must also remember to show gratitude for the personal talents that make our discoveries possible. Our intelligence, powers of reason, hard work, ingenuity, and perseverance are all gifts from God. We are "under obligation to him for what [we] know and enjoy: [we] are indebted to him for it all."[12]

Isaac Newton is a good example of someone who followed these guidelines and was enlightened by the Spirit. He set out to learn everything there was to know about mathematics in his day, thus preparing himself to receive new knowledge that led to new discoveries such as the invention of calculus and mathematical formulations describing the law of gravity. Notwithstanding his personal faults and occasional squabbles with other intellectuals, he exercised patience and humility in his scholarly pursuits. John Conduitt, a nephew-in-law who knew him personally, described Newton's life as "one continued series of labour, patience, humility, temperance, meekness, humanity, beneficence, & piety."[13] Lastly, evidence that he was thankful comes from his having uttered this now famous quote: "If I have seen further it is by standing on ye shoulders of giants."

Few of us will ever achieve the intellect of Isaac Newton in this life, yet as Latter-day Saints we are in a special position to receive knowledge from the Spirit of the Lord. Through His prophet Joseph Smith the Lord said:

> How long can rolling waters remain impure? What power shall stay the heavens? As well might man stretch forth his puny arm to stop the Missouri river in its decreed course, or to turn it up stream, as to hinder the Almighty from pouring down knowledge from heaven *upon the heads of the Latter-day Saints* (D&C 121:33; emphasis added).

The Lord is ever ready to pour down knowledge upon our heads. He has provided us with wonderful blessings unknown to previous generations to aid in this process. Latter-day revelation, scripture, and priesthood authority, to name a few, are invaluable assets in the quest for truth. We ought to lean on these blessings and prepare ourselves through study, prayer, and righteous living to receive the knowledge that the Lord is waiting to pour down upon our heads. When we receive knowledge, let us honor Him by sharing it with the world and bringing it to Zion so that others may be strengthened and edified.

May we always remember that the illuminating power of the Spirit of the Lord makes our Savior the ultimate source of truth and light. Without the guidance of His Spirit, humanity would flounder in ignorance and darkness. He truly is the Light of the World. As long as we follow after Him, He has promised, "Ye shall not walk in darkness" (John 8:12).

Notes

1. Thomas S. Monson, *Live the Good Life* (Salt Lake City: Deseret Book, 1988), 92.
2. As found in William L. Shirer, *The Rise and Fall of the Third Reich* (New York: Simon and Schuster, 1960), 986.
3. Royal L. Garff, *You Can Learn to Speak* (Salt Lake City: Deseret Book, 1979), 204.
4. John A. Widtsoe, *Evidences and Reconciliations* (Salt Lake City: Improvement Era), 178.
5. *Revelation, Reason, and Faith: Essays in Honor of Truman G. Madsen*. Edited by D. W. Parry, D. C. Peterson, and S. D. Ricks (Provo, Utah: FARMS, 2002), l.
6. Robert J. Matthews, *A Bible! A Bible!* (Salt Lake City: Bookcraft, 1990), 234—35.
7. Joseph Fielding Smith, Conference Report, October 1928, 101.
8. *Discourses of Brigham Young*. Selected and arranged by John A. Widtsoe (Salt Lake City: Deseret Book, 1954), 257.
9. Ibid., 250.
10. John A. Widtsoe, *In Search of Truth: Comments on the Gospel and Modern Thought* (Salt Lake City: Deseret Book, 1930), 114–15.
11. Ibid.
12. Brigham Young, *Discourses of Brigham Young*. Selected and arranged by John A. Widtsoe (Salt Lake City: Deseret Book, 1954), 2.
13. Richard Westfall, *The Life of Isaac Newton* (Boston, MA: Cambridge University Press, 1994), 306.

INDEX

A

Abraham 12–20, 31, 100, 102, 109, 113
Agency 55, 103, 108, 113–14, 118
Agnosticism vi, 87–89, 98–99, 108
Alexander (the Great) 32–33
Alexandria 32–35, 41, 45–46
Alma 8–9, 14, 24–25, 28, 84, 87–91, 103–104, 114, 127
Anaxagoras 27–28
Antirealism 98, 120
Apostasy x, 13, 15, 21, 26, 37, 40–42, 45–46, 50, 69
Aquinas, Thomas 43–45
Aristotelian 43, 44, 46, 56–57, 97
Aristotle ix, 26, 28, 30–32, 38, 40, 43–44, 49–50, 52, 83, 97
Arius 42
Arphaxadite 12
Assyria, Assyrian 18–19, 21
Astronomy 3, 14, 17–20, 23, 25–26, 46, 48–49, 54, 64–65
Atheism x, 27, 87–90, 99, 111
Athens 27–30, 38–39
Augustine 43–45

B

Babylon 12, 16, 20–22
Bacon, Roger 49–50
Benson, Ezra Taft 81–82, 88, 93
Bonaparte, Napoleon 88
Brigham Young University, BYU 4–5, 69, 133
Byzantine 48

C

Cajal, Santiago Ramón 72–73
Cannon, George Q. 5, 10
Chaldea 12, 14, 16–17, 19, 21–22, 25
Charlemagne 49
Cheops 18
Christosophy 43, 45

Columbus, Christopher 54, 69
Copernicus, Nicolaus x, 55–58, 70
Creation 13, 14, 26, 85, 87, 102, 109, 112
Creator 8, 67, 85, 88, 91, 113

D

Daniel 22–23, 27, 34, 123
Darius 33
Dark Ages x, 40, 43,–46, 48–50, 53
Darwin, Charles 87–88, 93
Dawkins, Richard 92
Deism, deist 85–86, 108
Descartes, René 55, 61–63, 68–70, 79, 84–85, 93, 108
Determinism 102–103, 117
Divine intervention 110–112

E

Egypt 12, 16–20, 25, 32–36, 41
Einstein, Albert 95–96, 105, 117, 120–121
Enlightenment 40, 126–127
Evolution viii, 87–88, 103

F

Faith viii, ix, 1, 8–9, 11, 19, 39, 42, 44, 60, 63, 73, 85, 88, 90–91, 104, 114, 126–127
First Illumination 30, 32, 40, 69
First Vision 4, 60
Flavius, Josephus 12, 16–17, 19, 24–25, 34–36
Fleming, Alexander 74–75
Flood 12–14

G

Galen, Claudius 47–49, 55
Galilei, Galileo x, 1, 9, 55–61, 63–64, 68–70, 79, 84–85, 93, 97–98, 105, 108
Golden Age v, 25, 32, 36

Gould, Steven J. 89, 108, 114
Greece, Greek ix, 3, 25–28, 32–36, 38–43, 46–52, 69
Gutenberg, Johann 54

H

Ham 16
Hammurabi 16, 21, 23–24
Heliocentrism 33, 56–58, 60
Herodotus 18, 24
Hinckley, Gordon B. 4
Homoousious 67

I

Idolatry 19–20, 23, 25, 27, 38, 39
Islam 48–49

J

Jerusalem 18–19, 25, 34–35, 117
Job 19, 31

K

Kekule, August 75–76, 81

L

Lehi 25, 113
Light of Christ vi–x, 7–9, 13–14, 25, 27–28, 3–32, 38, 40–41, 46, 50–51, 53–55, 61, 66, 69, 71–72, 81, 83, 84, 91–92, 96, 101, 112–114, 101, 117–118, 121, 123, 125–128
Ludlow, Daniel H. xi, 4, 9

M

Madsen, Truman 125, 128
Malaria 76
Manetho 18–19
Materialism 99–100, 120
McConkie, Bruce R. 24, 54, 69, 85, 93, 109, 114
Melchizedek 19
Mendel, Gregor 98–99
Mendeleyev, Dmitri 78–81

Mesopotamia v, 12, 14, 20–25
Middle Ages 46–47, 49, 51
Miletus 25–26, 39, 40
Mind-body problem 110–111
Monson, Thomas S. 124, 128
Mosiah 124

N

Nabopolassar 21–22
Naturalism 93, 99, 106, 114
Natural philosophy 26, 49
Nebuchadnezzar 22–23
Neoplatonism 43
Nephi 7, 42, 54, 60, 71, 86, 88–89, 91, 93, 103, 110, 113, 116–117, 124, 126
Neuron 71–73
Newton, Isaac x, 55, 64–70, 79, 84–85, 93, 96, 105, 108, 127–128
Nibley, Hugh 51, 115
Nicaea 42
Nihilism 6

O

Omar Bradley 125
Origen 41, 51

P

Packer, Boyd K. 100, 106, 121
Pascal, Blaise 91
Pasteur, Louis 71, 74
Paul (the Apostle) 8, 27, 38–40, 53, 69, 81, 84, 117, 123
Penfield, Wilder 104–106
Penicillin 74
Periodic Table vi, 72, 78–79
Philadelphus (Ptolemy II) 33–35
Philosophes 85
Philosophy viii, 5, 11, 26,–28, 32, 38, 40–43, 45, 48–49, 57, 61, 63, 90, 92, 100, 124, 133
Plato 28–30, 40, 41, 43
Popper, Karl 32, 36, 95, 105
Pratt, Orson 2, 4, 63
Pre-Socratic 25–28, 32
Priesthood 13

Problem of evil 110, 112
Process philosopher 120
Process theology 111–113
Ptolemies 32–36, 46–47
Ptolemy, Claudius 33–35, 45–47, 49, 55
Ptolemy II see Philadelphus
Pyramid 17–19

R

Reason 8–9, 26, 28, 30, 32, 40–41, 44, 63, 69, 85, 91–92, 97, 103–104, 127
Reductionism 99, 103–104
Relativism 6–7, 101
Renaissance 24, 36, 46, 50–51, 54
Restoration x, 4, 46, 51, 53–55, 61, 63–64, 68–69, 80–81
Roberts, B. H. 4
Ross, Ronald 76–77

S

Scholasticism 44, 50, 61, 63, 97
Science ix, 1, 3–5, 11, 17, 22, 25, 29, 32, 43, 49–50, 59, 61, 64–65, 69, 71, 86–89, 91–92, 95, 96, 98–99, 100–101, 105, 107–112, 114–115, 124–125, 133
Scientific Revolution 50, 53, 55, 108
Second Illumination x, 50, 64, 69, 84–85, 123
Secularism vii, viii, ix, x, 1– 5, 7–8, 11–13, 15–16, 20–23, 25, 32, 38–40, 43, 46, 48–49, 54–56, 66, 68–69, 71, 81, 85, 90–93, 108–109, 114, 123–125
Septuagint 35
Shem 12–13, 16
Shepherd-kings 18–19
Skepticism 6, 28, 50, 79, 118
Smith, Joseph vii, 3–5, 13, 17, 24, 60, 63, 68, 87, 127
Smith, Joseph F. ix, xi, 11, 87, 93
Smith, Joseph Fielding 13, 24, 37, 50, 52–53, 69, 126, 128
Socrates ix, 28–30, 38–39
Spirit see Light of Christ

Spirit of the Lord see Light of Chirst
Sumeria 12–16, 29–20

T

Talmage, James E. 4
Taylor, John 50, 52, 63
Technology 24, 36, 51, 96, 123
Telescope 2, 56–57, 74, 97
Thales ix, 25–26, 40
Theism 83–85, 99, 111
Truth vii, viii, ix, x, 1–9, 15, 21–23, 25, 27–28, 30–32, 38–39, 42–43, 45, 50, 54–56, 58–63, 66, 68–69, 73, 77–78, 81, 86–87, 93, 95, 98–101, 107–109, 114–115, 118, 123–128

U

University 4, 24, 43–44, 48, 51, 57, 65, 70, 78, 81, 93–94, 102, 105–106, 114, 116, 120–121, 128
Ur 12, 15, 16

V

Verisimilitude 7, 8
Vesalius 54
Victoria, Queen 91

W

Widtsoe, John A. xi, 4, 9–10, 21, 24, 42, 51, 93, 100, 105–106, 114, 117–118, 121, 125, 127–128
Word of Wisdom 2

X

Xenophanes 27–28

Y

Young, Brigham vii–ix, xi, 1–2, 4–6, 9–10, 96, 98, 105–106, 108–109, 114, 117, 121, 123, 126, 128

Z

Zion viii, ix, 2, 4–5, 114, 128

ABOUT THE AUTHOR

Dave Collingridge completed his PhD in the BYU Psychology Department's Theory and Philosophy program in 2003. He enjoys studying history and philosophy of science, and has published articles in peer-reviewed scientific journals. Dave has worked as an instructor for the BYU Salt Lake Center and Salt Lake Community College, and as a research statistician for Intermountain Healthcare. He lives in Utah with his wife and four children.